SOLID WASTES:

FACTORS INFLUENCING GENERATION RATES

DOUGLAS B. CARGO

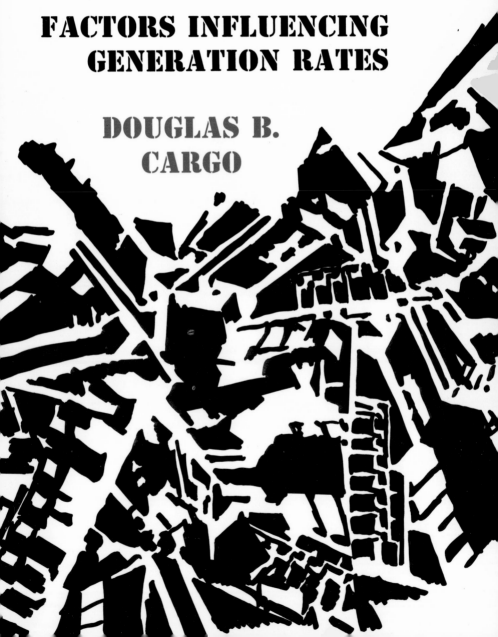

SOLID WASTES:

FACTORS INFLUENCING
GENERATION RATES

by

DOUGLAS B.
CARGO

The University of Texas at Dallas

THE UNIVERSITY OF CHICAGO
DEPARTMENT OF GEOGRAPHY
RESEARCH PAPER NO. 174

1978

To Veronica, Sean, Jason and Jonathan

Library of Congress Cataloging in Publication Data

Cargo, Douglas B., 1943 -
    Solid wastes: factors influencing generation rates.
    (Research Paper - University of Chicago, Dept. of Geography; no. 174)
    Bibliography: p. 99
    1.Refuse and Refuse Disposal - U.S. I. Title. II. Series: Chicago.
    University. Dept. of Geography. Research Paper; no. 174

H31.C514 no. 174 [TD788] 363.6
78-16823
ISBN 0/89065-081-0 paperback

TD
788
C37

Research Papers are available from:
The University of Chicago
Department of Geography
5828 S. University Avenue
Chicago, Illinois 60637
Price: $6.00 list; $5.00 series subscription

ACKNOWLEDGMENTS

I would like to acknowledge my indebtedness in the pre-
paration of this dissertation to many people. Special appreciation is owed to my advisor, Brian J. L. Berry. His encourage-
ment, understanding, and guidance helped this author overcome
many obstacles regarding graduate school, geography, academia
in general, and the dissertation. He has been, and will remain,
an inspiration.

Further acknowledgement is to be given to:

Sue Hull, Computer Assistant, University of Chicago,
Computer Center, for her untiring efforts at finding
my errors.

Susan Caris, Graduate Student, University of Chicago,
Department of Geography, for her patient coding and
plotting of thousands of case records.

Charles Butler, Environmental Protection Agency, Cincinnati,
Ohio, for his cooperation in locating access codes and
duplicating tapes only days before his job was ter-
minated.

David Morgan, Assistant Professor, University at Dallas,
for help in the design and interpretation of
statistical testing measures.

To my wife, Veronica, who endured hard times, long working
hours, and medical problems, while coping with three children
and a husband who was not always smiling, I extend this public
expression of appreciation.

To all these people, plus my parents, and fellow students
and faculty members of the Department of Geography as a whole:
thank you.

                                        Douglas B. Cargo
                                        Plano, Texas
                                        September, 1977

PREFACE

There has been only one attempt to survey the quantities of solid wastes produced in the United States as a whole and to evaluate solid waste disposal practices, the <u>1968 National Survey of Community Solid Waste Practices</u> conducted by the Bureau of Solid Waste Management of the U. S. Department of Health, Education, and Welfare. Once the data had been collected, federal fund restrictions prevented analysis.

Indeed, in 1973, the Office concerned with maintaining the data was closed by the Administration. A substantial body of information would have vanished if it had not been possible to obtain a set of the data at the University of Chicago. The purpose of the research reported herein was to analyze a portion of these data, focusing in particular on solid-waste generation rates.

D.B.C.

TABLE OF CONTENTS

Page

ACKNOWLEDGMENT   iii

PREFACE   iv

LIST OF TABLES   vii

LIST OF FIGURES   viii

CHAPTERS

   I. INTRODUCTION   1

       The Subject is Solid Waste   1
       The Purpose of the Research   4

  II. DEFINITIONAL PROBLEMS, BACKGROUND STUDIES AND SURVEYS   5

       Definitions   5
       Solid-Waste Data and Studies   7
       The National Survey of Community Solid-Waste Practices   14
       The Survey's Questionnaires   16
       The Data Set Used in This Research   17
       Published Data From the National Survey   26

 III. METHODS AND ANALYTICAL PROCEDURES   28

       The Solid-Waste Data   30
       The Sample and It's Distribution   31
       The Socio-Economic Data   34
       Solid-Waste Generation Rates: Their Development   37
       Computer Programs Used in Analysis   41

  IV. DATA ANALYSIS   44

       Data Characteristics   44
       Derived Solid-Waste Generation Rates   48

          Study Figures vs E.P.A. Figures   49
          Regional Variations   49
          State Generation Rates   51
          Density-Conversion-Factor Check   51
          Generation-Rate Check with 31 Virginia Cities   54

       Socio-Economic and Solid-Waste Generation Rate
       Correlates   56
       Multiple Linear Regression Models   56
       Regression Models Solved for Three Selected Cities   63

TABLE OF CONTENTS
(Cont'd)

Page

V.   CONCLUSIONS

Federal Government Involvement in Solid Waste   73
Purpose of Research   76
Specific Conclusions   77

APPENDIX: A. NATIONAL SURVEY OF COMMUNITY SOLID-WASTE PRACTICES:
               QUESTIONNAIRES   79

           B. LIST OF STATES AND COUNTIES FOR WHICH SOCIO-ECONOMIC
               DATA WAS COLLECTED   90

LITERATURE CITED

LIST OF TABLES

Table		Page
1.	Total Solid Waste Produced Annually in the United States	2
2.	Waste Classifications	6
3.	Solid-Waste Collection Systems of the United States	9
4.	Solid-Waste Generation by Dwelling type	9
5.	Municipal Solid-Waste-Collection Rates by Population Densities	12
6.	Solid-Waste Multipliers from the California Study	15
7.	Percent of Population Surveyed by the National Survey	21
8.	County Records by State	22
9.	Quality of Responses to Questions in the Community Description Report	25
10.	Number of Valid Case Records for the CDR	26
11.	Number of Sample Cases by State	35
12.	Solid-Waste Density, 1965	39
13.	Density of Various Solid-Waste Classifications	40
14.	Frequencies of Estimated 1968 Population	47
15.	Generation Rate Comparisons	50
16.	Generation Rates and Number of Cases by State of the Sample	52
17.	Density-Conversion-Factor Effect on Generation Rates	54
18.	Independent Cities of Virginia	55
19.	Correlation Coefficients	57
20.	Partial Regression Coefficients	59
21.	Predicted Generation Rates for Three Cities	71

LIST OF FIGURES

Figure   Page

1. A Solid – Waste View of the United States   3

2. Percentage of Population Surveyed (H.E.W. Region)   18

3. Spatial Coverage of Counties Contained in the LDR   19

4. Spatial Coverage of Counties in CDR   20

5. Case Sample Distribution by State   32

6. Spatial Coverage of Counties within the Sample   33

7. Histogram of Surveyed 1968 Population   46

8. Graphic Trends Between Independent and Dependent Variables   61

CHAPTER I

INTRODUCTION

THE SUBJECT IS SOLID WASTE

The subject of this dissertation is solid waste--unless collected, a more profound pollutant of land and particularly of cities than are effluent discharges into the air and water. Man has always had to contend with his wastes, but when he began to congregate in cities, waste and its disposal became a profound problem.

Because reasonably effective collection and disposal systems now exist in all our cities we too often fail to recognize that solid wastes are a pollution problem, except at disposal sites. But occasionally and quite dramatically events can change this, as when essential waste-collection services in both New York City and Baltimore were crippled by strikes in the summer of 1974. Within a very short period of time wastes began to overflow receptacles, scattered into the streets, and blocked pedestrian traffic on the sidewalks. Health officials became concerned as these effluents began to 'ripen' in the summer's heat. Only at times such as these do we really begin to realize the quantity of waste that each and every one of us produces, and the pollution problems that normally we have been able to keep under control.

Many large cities have indicated that they are very close to exhausting their remaining disposal space. There are, then, two problems: too much waste is being generated; and the supply of land available for the disposal of solid waste is being exhausted with relatively few possibilities for expansion. The sheer quantity of waste produced each year is indicated in

1970

Table 1. This table shows that the estimated amount of solid
waste produced by all waste sources in the United States is
over 3.5 billion tons per year. Figure 1 provides approxi-
mate visual interpretation of these estimates of solid waste
production by scaling the various states so that their area is
proportional to the amount of solid waste that each produces.
Because the disposal-site problem is more immediate, far more
attention has been given to it and to questions of recycling
and reuse, than has been given to the prior question of generation rates. Indeed, most national studies use quite crude
methods of estimating the amounts of wastes that are generated.

TABLE 1

TOTAL SOLID WASTE PRODUCED ANNUALLY IN THE UNITED STATES

Waste Sources	Tons Produced
Household, Commercial, Municipal Wastes	250 million tons/year
Industrial Wastes	110 million tons/year
Agricultural Wastes	550 million tons/year
Animal Wastes	1,500 million tons/year
Mineral Wastes	1,100 million tons/year
Total	3,510 million tons/year

3.5 BILLION TONS/YEAR

Source: Black, Ralph, J., The National Solid Wastes Survey:
An Interim Report, (Washington: U.S. Department of Health, Education,
and Welfare, Bureau of Solid Wastes Management, 1970), p. 48.

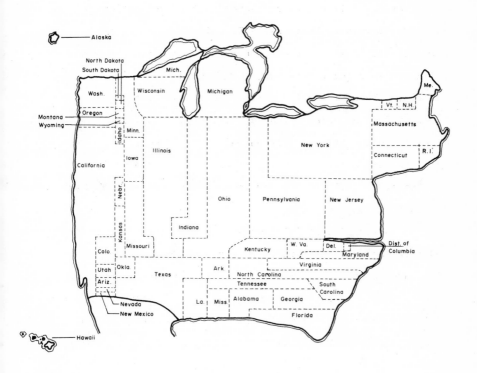

FIGURE 1

A SOLID-WASTE VIEW OF THE UNITED STATES

Source: Klee, Albert J., "Mapping the United States - A Solid
Waste View", Waste Age, September-October, 1970.

PURPOSE OF RESEARCH

Therefore, the purpose of the research reported in this study was to analyze data collected in the 1968 National Survey of Community Solid Waste Practices, a one-of-a-kind survey, to determine actual rather than estimated amounts of solid wastes being generated, to determine the variations existing between the rates of solid-waste generation by source, to determine the effects that socio-economic variables have upon the solid-waste generation rates, and whether there are any regional variations in the rates of waste generation, thus attempting to fill a major gap in knowledge relating to environmental pollution.

Chapter II deals with essential matters of definition and looks at existing methods of estimating solid-waste generation rates. Then Chapter III discusses sampling methods and the analytical procedures used. Chapter IV is devoted to data analysis. Finally, the principal conclusions are drawn together in Chapter V.

CHAPTER II

DEFINITIONAL PROBLEMS AND BACKGROUND STUDIES

DEFINITIONS

Wastes vary in size, form, origin, and physical composi-
tion. They are often placed into three categories; solid,
liquid and gaseous. Table 2 indicates the variety of wastes
conventionally considered under each of the above headings.
While this table is not all-inclusive, it does provide a good
cross-section of the dimensions of the national effluent
problem.

Solid Waste is defined by the United States Environmental
Protection Agency as "useless, unwanted, or discarded material
with insufficient liquid content to be free flowing."[1] It is
also common to classify solid-waste materials by their origins,
usually into those derived from domestic, municipal, commercial,
and industrial sources, viz:[2]

Domestic Refuse: includes all those types which normally
originate in the residential household or apartment house.

Municipal Refuse: embraces all the types which originate
on municipally owned property. These include street
sweepings and liter, catch-basin dirt, refuse from parks,
playgrounds, zoos, schools and other institutional
buildings, plus: solid wastes from sewerage systems.

---

[1]Solid Waste Management Glossary, (Washington: U.S.
Environmental Protection Agency, 1972), p. 17.

[2]Proceeding National Conference on Solid Wastes Research
(Chicago: Research Foundation, American Public Works
Association, 1963), p. 18.

TABLE 2

WASTE CLASSIFICATIONS

I. Solid Wastes
  A. Putrescibles
    Household Garbage
    Vegetable and Fruit
      Processing Wastes
    Animal Manure
    Dead Animals
    Meat, Poultry and Seafood
      Processing Wastes
    Others, Not Elsewhere
      Classified
  B. Bulky Combustibles
    Wood
    Paper and Paper Products
    Cloth
    Plastics
    Rubber
    Leather
    Yard and Street Wastes
  C. Bulky Non-Combustibles
    Metals
    Minerals
  D. Small Combustibles
    Wood
    Paper and Paper Products
    Cloth
    Plastics
    Rubber
    Leather
    Yard and Street Wastes
  E. Small Non-Combustibles
    Metal
    Mineral
    Ashes
  F. Non-Empty Cans, Bottles and
    Drums
  G. Gas Cylinders
  H. Powders and Dusts
    Organic
    Metallic Inorganic
    Non-Metallic Inorganic
    Explosive
  I. Pathological Wastes
    Cloth, Paper and Plastic
    Animal and Human Wastes
    Instruments and Utensils
  J. Sludges
    Chlorinated

I. Solid Wastes (cont.)
  J. Sludges (cont.)
    Brominated
    Fluorinated
    Acid
    Alkaline
    Water-Reactive (Unhydro-
      lyzed)
    Air-Reactive
    Putrescible
    Miscellaneous Organic
    Metallic Inorganic
    Non-Metallic Inorganic
  K. Demolition and Construction
  L. Abandoned Vehicles
  M. Radiological Wastes

II. Liquid Wastes
  A. Wastewaters
  B. Contaminated Waters
    Chlorinated
    Brominated
    Fluorinated
    Acid
    Alkaline
    Putrescibles
    Insoluble Oils
    Soluble Oils
    Toxic Organics
    Toxic Inorganics
    Soluble Metals
    Others, NEC
  C. Liquid Organics
    Chlorinated, Brominated
    Fluorinated, Sulfurated
    Acid, Alkaline
    Water-Reactive (Unhydro-
      lyzed)
    Shock-Reactive
    Soluble Metals
    Others, NEC
  D. Tars
  E. Slurries

III. Gaseous Wastes
  A. Odorous
  B. Particulate Combustibles
  C. Organic Vapors, Acid Gases

Source: Weston, Roy F., New York Solid Waste Management Plan, Status Report 1970. (Washington: United States Environmental Protection Agency 1971), Table 17-1, (SW-5tsg).

Industrial Refuse: includes all solid waste which result from industrial processes and manufacturing operations such as factories, processing plants, repair and
cleaning establishments, refineries and rendering
plants.

Additional waste-origin categories sometimes used are:[1]

Agricultural - resulting from rearing or slaughtering
of animals, and orchards and field crops

Institutional - originating from educational, health
care, and research facilities

Pesticides - residue from manufacturing, handling or
use of chemical for industry, plant, and animal

Residential - originates in a residential environment

Hazardous - radioactive wastes

It was this EPA-approved classification that was used by the
National Survey of Community Solid Waste Practices, the data
from which is analyzed in the chapters that follow.

SOLID-WASTE DATA AND STUDIES

This survey was undertaken because very few communities
ever measure their solid wastes, let alone provide detailed
source-oriented statistics. Where measurements are made, there
is confusion about the relevant units -- pounds, tons, cubic
yards, truck loads, etc. -- and about exactly what should be
included within these measurements -- publicly collected,

_____

[1]Solid Waste Management Glossary, p. 17

wastes disposed of privately, or even those wastes which individuals haul to disposal sites on their own. Table 3 indicates
the relative responsibility of each collection system for three
specific waste sources. Public collection systems take care
of the majority of household wastes, private systems care for
the majority of commercial wastes and private haulers accommodate the majority of the industrial wastes.

At the present, most localities have no mandates that
require any or all of these systems to keep quantitative data.
There is, as a result, very little known about variations in
solid-waste generation rates, or the reasons for the variations, and such information that is available is frequently
counter-intuitive.

For example, a survey by Davidson[1] of low-income residential areas in Cincinnati, Ohio, indicated that the amounts of
solid waste generated per day per capita were much higher than
the generation rates for the state on the average, and that
apartment dwellers produce about half of the waste that a single
family residence does (see Table 4). Davidson's study also
indicated that the average generation rate by dwelling type
decreases as the number of persons per dwelling unit increases.
Finally, he concluded that the quantity of solid wastes generated from a dwelling unit depends upon the number of occupants,
not the dwelling unit, that a fixed amount of solid waste
(junk mail, lawn trimmings, etc.) is generated per dwelling

---

[1]George R. Davidson, Residential Solid Waste Generation in Low
   Income Areas (Washington: U.S. Environmental Protection
   Agency, 1972).

TABLE 3

SOLID-WASTE COLLECTION SYSTEMS OF THE UNITED STATES

	Public	Private	Individual
Household Wastes	56%	32%	12%
Commercial Wastes	25%	62%	13%
Industrial Wastes	13%	57%	30%

Source: Black, Ralph J., <u>The National Survey of Community Solid Waste Practices: An Interim Report</u>. (Washington: U.S. Department of Health, Education and Welfare, Bureau of Solid Wastes Management, 1970), p. 10.

TABLE 4

SOLID WASTE GENERATION BY DWELLING TYPE

	Single Family Housing Unit	Multiple Family Housing Unit	Apartment House
Pounds/capita/week	12.54	9.83	6.91
Gallons/capita/week	14.07	11.00	5.61
Pounds/cubic yard	179.97	180.50	248.87

Source: Davidson, George R., <u>Residential Solid Waste Generation in Low Income Areas</u>, (Washington: United States Environmental Protection Agency, 1972), p. 2.

unit regardless of the number of persons occupying it, and at the same time that the average solid waste contributed per person is constant within each of the classes of dwelling units studied (single-family, multi-family, and apartment).

While similar types of studies for other income levels at different residential locations have not been undertaken, the study by Davidson suggests that the results of such studies may vary dramatically from what is today often taken to be common knowledge, i.e., that greater wealth automatically implies more solid wastes and indeed, that the most fundamental relationship is the mix between land use and the affluence of population. One thing that remains unclear is the relationship of population density to generation rates. The New York State Plan[1] indicated the following:

1. The more densely populated municipalities are able to offer better municipal services, such as increased frequency of collection of refuse. There is a tendency of households both to generate larger amounts of wastes with increased frequency of refuse collection and also to rely less on on-site disposal such as backyard burning.

2. The more densely populated urban areas also are able to support more institutional and commercial activities per capita, resulting in an increased per capita production of municipal waste. Regionalized shopping may also add to this (situation).

3. Urban renewal, a major source of demolition and construction debris is usually more extensive in the larger urban centers than in the more rural communities..

Furthermore, the same study indicated that as population and population density increases so does the amount of solid

---

[1] New York Solid Waste Management Plan, Status Report 1970, (Washington: U. S. Environmental Protection Agency, 1971).

waste generated per capita. But this increase is not limitless
and the increase in solid waste seems to level off as density
reaches a range of about 8,000-10,000 persons per square mile
(see Table 5). As the authors of the New York State Plan
reported: "The general trend in per capita collection is to
increase up to a certain density and then decrease. The apparent range for this transition is around 8,000-10,000 people per
square mile, which is also the density range generally marking
the transition from single-family dwelling communities to
multi-family dwelling communities."[1]

There is better information about manufacturing wastes.
Two reports by Gene Steiker[2] and Thomas Knotuly[3] dealt with
solid-waste generation coefficients in the manufacturing sectors
of the economy. They calculated solid-waste coefficients as
weighted averages as derived from a questionnaire survey undertaken in the Philadelphia metropolitan region by the Regional
Science Research Institute. The data were assembled at the
3rd and 4th digit levels of the Standard Industrial Classification Code. The end result was that the authors arrived at an
understanding of waste generation in tons per employee per year
for 14 manufacturing materials and 396 SIC 4-digit industries.
In addition generation rates were developed for 9 non-manufac-
turing sectors of the economy.

---

[1] Ibid., p. 13-14

[2] Gene Steiker, Solid Waste Generation Coefficients: Manufacturing Sectors, (Philadelphia: Regional Science Research
Institute, Working Paper, April, 1974).

[3] Thomas Knotuly, Solid Waste Generation Coefficients: Non-
Manufacturing Sectors, (Philadelphia: Regional Science Research
Institute, Working Paper, April, 1974).

TABLE 5

MUNICIPAL SOLID-WASTE COLLECTION RATES

BY POPULATION DENSITIES

(pounds/capita/day)

Population	Waste Collected by Pop. Density Ranges		
	0-399/sq.mi	4000-6999/sq.mi	7000 +/sq.mi
0-4999	3.3	--	--
5000-19999	3.6	5.0	4.6
20000-99999	4.1	4.1	4.6
1000000 +	4.6	5.1	5.6

Source: New York Solid Waste Management Plan, Status Report 1970, (Washington: United States Environmental Protection Agency, 1971), p. 3-11.

When the authors turned to residential and commercial wastes, sound information was unavailable. Thus, it was "assumed that each person in the Philadelphia SMSA generates 3.5 pounds of household wastes per day,"[1] an assumption that is questionable.

Because of such inadequacies in the available information base, it has become necessary for governmental and planning agencies to devise indirect methods of predicting the amounts of solid waste generated by households and commercial activities, to aid in solid-waste management and the development and implementation of long-range planning goals. The method used most commonly is very crude and inaccurate: the population ratio model. This method simply divides the total amount of solid waste estimated to have been produced by the total population producing it to yield a 'generation rate.' The rate is a figure which can then be applied to other populations to allow solid-waste generation to be estimated. One inaccuracy in this method lies in the total-waste statistic used. The fact is there are few, if any, accurate total-waste figures available for any given geographic or politically-bounded area. Another problem is that waste generation may vary from one type of community to another, so that even a good overall ratio may be a poor prediction in any given specific case.

To attempt to offset these inadequacies of the single generation-rate method, a more comprehensive and innovative

---

[1] Ibid., p. 2.

analysis of solid-waste generation was conducted by the State of California.[1] In this study, a complete county and disposal site analysis of sources and totals of solid waste was undertaken. The amounts of wastes by source units, by city size, and per capita, were calculated for all of the major solid waste categories (ie. commercial, industrial, etc.). These calculations allowed the California staff to arrive at waste totals and waste production on a county level.

The most noteworthy aspect of the study was the development of waste multipliers by specific source units, made possible because the investigation collated specific and concise data from both public and private sources.

The multipliers were developed on a per-capita basis for such sources as residences, on a square-mile basis for street dirt, on an employee basis for many industries acccording to the Standard Industrial Classification, and on a city-size basis. By developing these alternative multipliers the California staff were able to estimate the amount of waste generated by a certain area when reliable solid-waste information was unavailable from either public or private sources. Some of these waste multipliers are summarized in Table 6.

THE NATIONAL SURVEY OF COMMUNITY SOLID-WASTE PRACTICES

The California study clearly indicated the need for an input data base upon which proper waste analysis might be computed. EPA attempted to develop such a data source in the National Survey of Community Solid Waste Practices. This

---

[1] California Solid Waste Management Study 1968 and Plan 1970, (Washington: U. S. Environmental Protection Agency, 1971).

TABLE 6

SOLID-WASTE MULTIPLIERS FROM THE CALIFORNIA STUDY

Waste Source	Annual wastes Tons/employee
Commercial and Public Facilities	3.81
Demolition and Construction	41.25
Industries (Selected)	
Canning and Preserving	.5.56
Lumber and Wood Products	21.69
Textiles	0.52
Primary Metals	6.73
Printing and Publishing	13.20
Stone Clay and Glass, Concrete	18.11
Residential	
Single Family	1.43 tons/unit
Multiple Family	0.63 tons/unit
City Streets, Tree Trimmings, Lawns	
Leaves	42.9 lb/capita/year
Sewerage Treatment	87.1 lb/capita/year
Local Park	5.4 lb capita/year
Freeway Refuse	8.0 lb/capita/year

Alternate Multipliers

Populations & Services	Commercial & Services	Demolition & Construction
100,000	3.5 lb/capita/day	500 lb/capita/day
10,001-100,000	2.5 lb/capita/day	250 lb/capita/day
1,001-10,000	2.0 lb/capita/day	100 lb/capita/day

Source: Compiled from tables in California Solid Waste Management Study 1968, and Plan 1970, (Washington: United States Environmental Protection Agency, 1971).

survey was undertaken as a consequence of a recommendation made in July, 1966, at the Conference of Sanitary Engineers, to the effect that the Solid-Waste Program of the U. S. Department of Health, Education, and Welfare, create a list of essential data and submit guidelines for the conducting of statewide surveys. From this emerged The National Survey of Community Solid Waste Practices. The actual forms and specifications of the survey were developed with the help of outside consultants, state agencies and the Solid-Waste Program of the Department of Health, Education and Welfare.

THE SURVEY'S QUESTIONNAIRE

The 1968 National Survey was composed of three questionnaires: the Community Description Report (CDR), the Land Disposal Site Report (LDR), and the Facility Investigation Report (FIR). (See Appendix A)

The Community Description Report questionnaire covered four broad information categories:[1]

1. storage
2. collection
3. disposal
4. budget and fiscal

The Land Disposal Report and the Facility Investigation Report questionnaires covered three general areas of information:[1]

1. description and evaluation of the site
2. quantitative data
3. fiscal data

The survey forms were delivered to the various states which, at that time were receiving the Federal solid-waste planning grants. Seminars and workshops were conducted by the

---

[1]Ralph J. Black, et al., The National Solid Waste Survey: An Interim Report. (Washington: U.S. Department of Health, Education, and Welfare, Bureau of Solid Waste Management, 1970), p. 2.

[2]Ibid, p.3

Solid-Waste Program with the various state personnel. It was
thought that "by using the personal interview technique and by
providing sufficient guidance to the interviewer through the
instruction manual and seminar presentation...the National
Survey data would prove both uniform and reliable."[1]

The implementation of the survey was carried out by
the state agency receiving the planning grant. According to
the Preliminary Data Analysis, "to obtain reliable data, the
Survey was performed as a field investigation, with individuals
or teams of data collectors actually visiting the communities
and sites. Information was obtained either by personal interview or direct observation; under no circumstances was a form
completed on a mail-out-and-return basis."[2] Once completed the
forms were returned by the states to the Solid-Waste Program.

THE DATA SET USED IN THIS RESEARCH

The data set used in this research initially consisted
of information from two of the three questionnaires used in the
National Survey, the Community Description Report and the Land
Disposal Site Report (See Appendix A for a copy of the complete
questionnaire from all three questionnaires used in the survey).

The Community Description Report is composed of 12,142
records. Each record contains data from the 40 questions of
the questionnaires. The Land Disposal Report contains 15,679
case records. The Facility Investigation was not obtained for
this research; however a copy of that questionnaire is also
contained in Appendix A.

---

[1] A. F. Muhich, A. J. Klee, and P. W. Britton, Preliminary Data
Analysis: (Cincinnati: U. S. Dept. of HEW, Public Health
Service, Publication #1867, 1968), p. vii.

[2] Ibid, p. vii.

The entire survey covered solid waste data from 46% of
the nation's population in 1967. Seventy-five percent of this
population was considered to be urban and living in over 6,259
communities.[1]

Figure 2 indicates the percentage of the population
covered, by HEW regions. The highest percentage of the population surveyed was on the West Coast with the mid-Atlantic
states second. The percentages for individual states are
summarized in Table 7. Of the 34 states, the State of Washington and the District of Columbia had 100 percent coverage.

The areal coverage of the two questionnaires used in
this research varied. The Land Disposal Site Report covered
all states except Wisconsin and Nebraska. Figure 3 indicates
this coverage. Those counties which had at least one data
record for the survey are shaded.

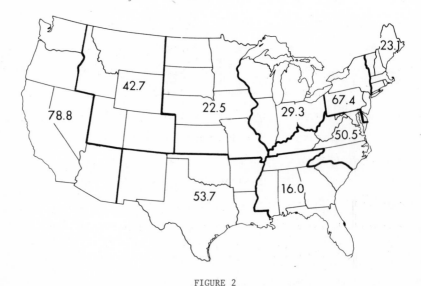

FIGURE 2

[1]Black, p. 7.

PERCENTAGE OF POPULATION SURVEYED
(HEW REGIONS)

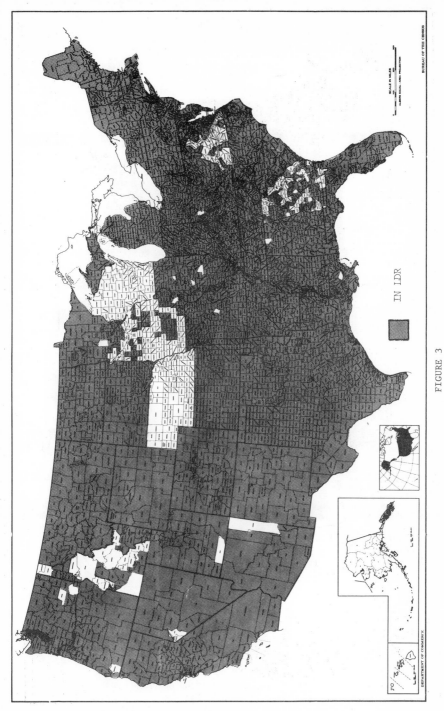

FIGURE 3

SPATIAL COVERAGE OF COUNTIES CONTAINED IN THE LDR

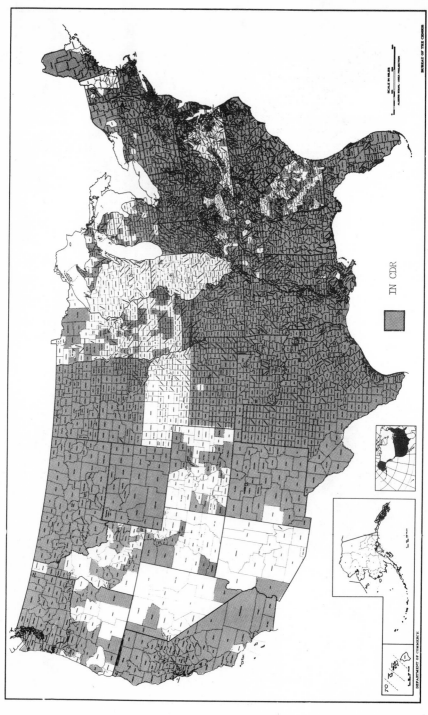

IN CDR

FIGURE 4

TABLE 7

PERCENT OF POPULATION SURVEYED BY STATE

BY THE NATIONAL SURVEY

State	Percent Population Surveyed	State	Percent Population Surveyed
1. Arkansas	30.3	18. Montana	65.6
2. California	94.9	19. New Jersey . . .	40.8
3. Colorado	59.7	20. New Mexico . . .	91.4
4. Connecticut	87.3	21. New York	58.3
5. Delaware	79.4	22. North Carolina .	44.0
6. District of Columbia .	100.0	23. North Dakota . .	96.0
7. Florida	8.3	24. Ohio	58.9
8. Georgia	36.6	25. Oklahoma	61.3
9. Hawaii	30.7	26. Oregon	3.1
10. Idaho	52.9	27. Pennsylvania . .	94.8
11. *Indiana	64.2	28. Rhode Island . .	24.8
12. *Iowa	58.3	29. South Carolina .	53.0
13. *Kansas	57.3	30. Tennessee	3.5
14. Kentucky	68.1	31. Texas	41.3
15. Louisiana	86.8	32. Virginia	64.0
16. Maryland	24.1	33. Washington . . .	100.0
17. Michigan	24.2	34. West Virginia . .	96.9

*   These states were surveyed by
    personnel of the Solid Wastes
    Program, with the cooperation
    of the appropriate state agen-
    cies, to attain an adequate
    sample for the Regions involved.

Source: A.J. Muhich, A.J. Klee and P.W. Britton, Preliminary Data
          Analysis: 1968 National Survey of Community Solid Waste
          Practices, (Cincinnati: United States Department of Health,
          Education and Welfare, Public Health Service, 1968), p. ix.

TABLE 8

COUNTY RECORDS BY STATE

State	Total Records/State	Total Counties	Counties With Records	Average Record/County
Alabama	331	67	67	4.9
Arizona	1	14	1	1.0
Arkansas	150	75	75	2.0
California	449	58	58	7.7
Colorado	29	63	20	1.4
Connecticut	161	8	8	20.1
Delaware	42	3	3	14.0
Washington, D.C.	1	-	-	1.0
Florida	198	67	67	2.9
Georgia	160	159	90	1.7
Idaho	43	44	23	1.8
Illinois	19	102	1	19.0
Indiana	196	92	92	2.1
Iowa	96	99	41	2.3
Kansas	178	105	103	1.7
Kentucky	100	120	80	1.2
Louisiana	244	64	64	3.8
Maine	38	16	12	3.1
Maryland	77	23	23	3.3
Massachusetts	329	14	14	23.5
Michigan	459	83	53	8.6
Minnesota	39	87	33	1.1
Mississippi	397	82	81	4.9
Missouri	460	115	115	4.0
Montana	109	56	54	2.0
Nebraska	-	-	-	-
Nevada	9	17	3	3.0
New Hampshire	-	-	-	-
New Jersey	566	21	21	26.0
New Mexico	94	32	32	2.9
New York	583	62	60	9.7
North Carolina	249	100	97	2.5
North Dakota	283	53	53	5.3
Ohio	919	88	88	10.4
Oklahoma	195	77	77	2.5
Oregon	67	36	19	3.5
Pennsylvania	2,665	67	67	39.7
Rhode Island	39	5	5	7.8
South Carolina	278	46	46	6.0
South Dakota	231	67	67	3.4
Tennessee	397	95	95	3.9
Texas	767	254	254	3.0
Utah	33	29	10	3.3
Vermont	24	14	10	2.4
Virginia	44	96	22	2.0
Washington	257	39	35	7.3
West Virginia	191	55	55	3.4
Wisconsin	-	-	-	-
Wyoming	66	23	23	2.8
Alaska	-	-		
Hawaii		5		

The Community Description Report did not cover as much area as the LDR. Initially only 34 states were surveyed, but additional data were added over the years so that ultimately only the states of Nebraska, New Hampshire, Alaska and Wisconsin were missing (see Figure 4). As with the LDR map, those counties with at least one record were shaded. That significant data coverage gaps exist is evident, and this conclusion is confirmed in Table 8, which shows the total number of counties for each state, the total number of case records and the total number of state records. The average number of determined case records per county is small.

Unfortunately, there were many problems with the way in which the National Survey Staff coded and stored their data. For example, as we began our search we made an initial sequential file listing of the LDR. It soon became apparent that a critical piece of data was missing that precluded us from making any kind of comparison between the LDR and CDR. This was the percent of political jurisdiction served by the disposal site. Without this datum in each case, no comparison could be made between the amount of waste generated and the amount of waste disposed. Thus, the data from the entire LDR questionnaire was eliminated from the research.

What type of waste data did the CDR provide? Question twenty of the CDR asked for "amounts (measured and estimated) of community solid wastes collected annually."[1] The respondent

---

[1] Muhich, p. xiv.

was provided with several waste source categories to which he could assign the data and then was asked to indicate the weights or volumes of the waste categories. Those waste source categories were: refuse (household, commercial, combined, industrial, institutional, agricultural), demolition and construction, street and alley cleaning, tree and landscape refuse, park and beach refuse, catch basin refuse, and sewerage treatment plant solids and pumping station cleanings. (See question twenty of the CDR in Appendix A)

The EPA indicates that the answers to these categories of waste sources are 'poor'. In fact, they estimate that 56.2% of the answers were poor. A quick review of Table 9 indicates that many of the answers to many other questions can be similarly rated.

While the quality of the answers was poor, it must again be emphasized that this was a 'one-of-a-kind' survey. At no other time in history has an attempt been made to collect a uniform body of waste data for the entire nation. For this reason, therefore, the data in the survey still are significantly better than having no data at all.

A decision was made to include in the study of the National Survey only the data which pertained to household, commercial and combined waste sources. Thus, the core of waste data used in the research reflected in this study is as follows:

Household Waste: (a) estimated where the community did not measure its waste
(b) actually measured by the community

Commercial Waste: (a) estimated
(b) measured

Combined Waste: (a) estimated
(b) measured

TABLE 9

QUALITY OF RESPONSES TO QUESTIONS IN THE
COMMUNITY DESCRIPTION REPORT

Question	"Good" (%)	"Fair" (%)	"Poor" (%)	Question	"Good" (%)	"Fair" (%)	"Poor" (%)
7	100.0	0	0	22	87.1	12.9	0
8	46.7	53.3	0	23	75.0	25.0	0
9	78.7	17.8	3.5	24	81.3	18.7	0
10	56.0	32.0	12.0	25	71.9	25.0	3.1
11	78.1	18.8	3.1	26	93.8	6.2	0
12	75.0	21.9	3.1	27	96.4	3.6	0
13	87.5	12.5	0	28	46.9	53.1	0
14	12.5	37.5	50.0	29	9.4	46.9	43.7
15	68.7	25.0	6.3	30	6.3	6.2	87.5
16	71.8	21.9	6.3	31	6.2	9.4	84.4
17	9.3	31.3	59.4	32	12.5	21.9	65.6
18	16.6	26.7	56.7	33	25.1	15.6	59.3
19	21.9	59.3	18.8	34	28.2	62.5	9.3
20	6.3	37.5	56.2	35	28.1	59.4	12.5
21	70.4	25.9	3.7	36	28.1	53.1	18.8

Source: A.J. Muhich, A.J. Klee and P.W. Britton, Preliminary Data Analysis: 1968 National Survey of Community Solid Waste Practices, (Cincinnati: United States Department of Health, Education and Welfare, Public Health Service, 1968), p. x.

The household waste source is restricted to waste from residential sources, while commercial waste includes waste from retail and wholesale establishments. It does not include waste from institutional or office complexes. The combined waste source category includes both the commercial and household waste. Where a community recorded waste by source, it provided information under the household and commercial categories; communities collecting in an undifferentiated manner reported only combined waste data. Table 10 indicates the number of case records available in the CDR for each of the waste source categories. For each of the waste source categories, the survey also made provisions for specification of whether or not the assigned value was in measurement units of tons or cubic yards.

TABLE 10

NUMBER OF VALID CASE RECORDS FOR THE CDR

Waste Source	Valid Cases*	% Responding to Individual Cases
Household (estimated)	547	5%
Household (measured)	3425	28%
Total Responding	3972	33%
Commercial (estimated)	527	4%
Commercial (measured)	2903	24%
Total Responding	3430	28%
Combined Household & Commercial (estimated)	775	6%
Combined Household & Commercial (measured)	7407	61%
	8182	67%

* Valid Cases are defined as those records having at least one waste source value.

Total number of Record Cases for the CDR, 12,142

PUBLISHED DATA FROM THE NATIONAL SURVEY

Only three publications have been published with data from or concerning the 1968 National Survey of Community Solid Waste Practices. These are: The National Solid Waste Survey: An Interim Report[1], Preliminary Data Analysis: 1968

[1]Ralph J. Black, Anton J. Muhich et al., The National Solid Waste Survey: An Interim Report, (Washington: U. S. Dept. of HEW, Public Health Service, 1970).

National Survey of Community Solid Waste Practices[1], and the
1968 National Survey of Community Solid Waste Practices:
Regions 1 & 2.[2] These three publications revealed little in
the way of analysis of the survey's data. The purpose of this
study is, therefore, to analyze the data from the National Survey of
Community Solid Waste Practices, especially to elicit good
summary information about generation rates and their deter-
minants.

---

[1]A. J. Muhich, A. J. Klee, and P. W. Britton, Preliminary Data
Analysis: 1968 National Survey of Community Solid Waste
Practices, (Washington: U. S. Dept. of HEW, Public Health
Service Publication #1867, 1968).

[2]Anton J. Muhich, Albert J. Klee, Charles R. Hampel, The 1968
Survey of Community Solid Waste Practices, Regions 1 & 2,
(Washington: U. S. Dept. of HEW, Public Health Service, 1969).

CHAPTER III

METHODS AND ANALYTICAL PROCEDURES

Initial inquiries to the Solid Waste Management Office of the Environmental Protection Agency were made in December of 1972, with the hopes of obtaining a duplicate set of computer tapes which contained the complete data from the 1968 National Survey of Community Solid Waste Practices. After many months devoted to locating the tapes, obtaining access to the data, and finding the correct location codes, a set of tapes was sent to the Cincinnati, Ohio, office for duplication. They were received back in Chicago in August of 1973. Along with the tapes were several documents which helped explain the coding sequence of the data. Among these documents was the Coding Manual for the National Survey of Community Solid Waste Practices[1].

Upon arrival the tapes were taken to the University of Chicago's Computation Center. A complete sequential file dump of both the CDR and the LDR was obtained. Each of these was found to be in a hexadecimal format. Only the location codes for each case record of the sequential file dump, for both the CDR and LDR, were in a decimal format. This allowed the location and identification of each case record.

Identification of each case record of the sequential file dump was made with the use of a government publication

---

[1] Coding Manual: The National Survey of Community Solid Waste Practices, (Washington U. S. Public Health Service, Solid Waste Program, September, 1967).

entitled <u>Geographical Location Codes</u>[1]. This establishes a set
of standardized codes for geographical locations throughout the
United States, and indeed the world. This was used exclusively
by the National Survey.

The location codes for the United States included codes
for each state, each county within each state, and most communities within each county of each state. The state code was
two digits (i.e., 01 for Alabama), the county three (i.e., 003),
and the community codes which were four digits (i.e., 2596).

These location codes were the first 12 characters (i.e.,
digits) of each record for both the CDR and LDR, and as such
could easily be interpreted from the code manual.

After decoding and identifying each of the 12,142
records of the CDR and the 15,679 records of the LDR, each was
mapped at the state and county levels. This was necessary, as
the spatial dimensions of the survey were not known.

Because the data on the duplicate tapes were in a hexadecimal format (packed decimal format, two digits per byte) a
COBOL program was used to "unpack" the data (i.e., convert the
data to an external decimal format--one digit per byte) to
facilitate using the data within various analysis programs.
This was necessary, since most computerprograms do not allow for
the input of packed decimal data.

---

[1]<u>Geographical Location Codes</u>, (Washington: U. S. Government
Printing Office, General Services Administration, Office of
Finance, October, 1966), (FSS stock Number 7610-926-9078).

THE SOLID-WASTE DATA

Once the conversion from the hexadecimal format to a decimal format was accomplished, another sequential file dump of each of the data sets was made. This time however, only selected data were included. For the CDR these included the following:

Location Codes

Population 1960

Population Estimated for 1968

Household Waste (estimated)

Household Waste (measured)

Household Waste in Tons or Cubic Yards

Commercial Waste (estimated)

Commercial Waste (measured)

Commercial Waste in Tons or Cubic Yards

Combined Waste (estimated)

Combined Waste (measured)

Combined Waste in Tons or Cubic Yards

These data were then printed out for each of the 12,142 records of the Community Description Report.

Likewise after the conversion for the LDR, a sequential file dump of selected data was made. It was at this point that the exclusion of the jurisdictional data from the LDR was discovered. Nevertheless the sequential file dump was made because of possible future reference.

The data contained on the Land Disposal Site Report Data dump included the following:

Location Codes

Zoning Regulations Enforced

Land Use Around Disposal Site

Number of Loads Daily Public

Number of Loads Daily Private

Number of Loads Daily Other

Quantitative Records

Tons Received (Weighted)

Tons Received (Estimated)

Cubic Yards Received

These data were obtained for each of the 15,679 records of the LDR. It should be recalled, however, that the entire data set from the LDR was eliminated from the actual research due to the lack of data which would have allowed comparisons of the data sets.

THE SAMPLE AND ITS DISTRIBUTION

Early in the research it was apparent that a sample of the waste data was necessary. The sample was to be taken of the 12,142 case records of the CDR.

It was decided that a workable number of sample cases would be in the range of 300-500, given constraints of time and funds available. Several sample selections were made and the spatial coverage of the samples checked. A systematic sample of one-thirtieth (1/30) of the 12,142 case records provided a workable data set. Starting the systematic sampling with case number one, 1/30 sample of the 12,142 cases resulted in a sample size of 405 cases. A review of these 405 cases revealed that several were actually 'state headers' or introductions to the particular state data. These 'headers' were eliminated so that the final sample had 397 case records.

Of the 397 case records of the sample, 43 states are represented. This is illustrated by Figure 5 which shows the number of sample case records per state. Of these 43 states the maximum number of case records per state was 84 and the minimum was 1. The number of counties which the sample represent is 359. This is further illustrated by Figure 6 in which those counties which are represented on the sample by at least one case record are shaded.

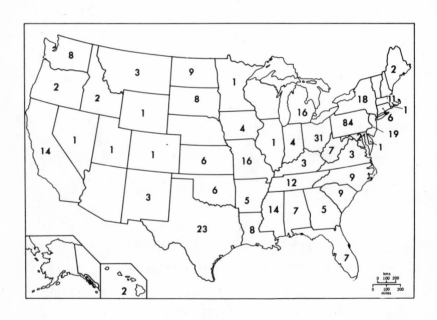

FIGURE 5

CASE SAMPLE DISTRIBUTION BY STATE

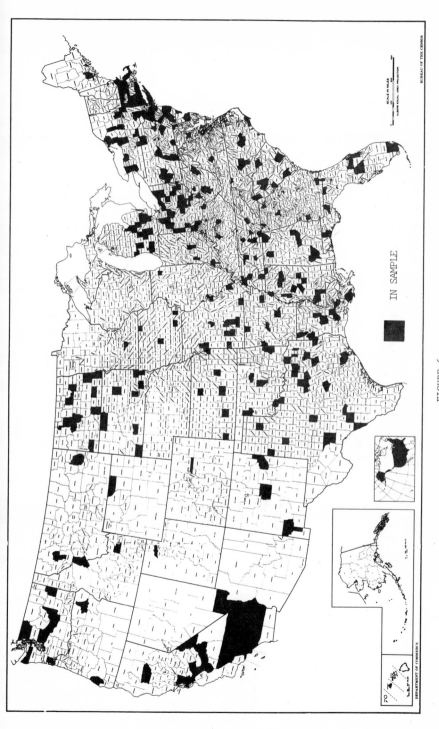

FIGURE 6

SPATIAL COVERAGE OF COUNTIES WITHIN THE SAMPLE

Table 11 indicates the number of cases per state and the percentage that those cases represent of the sample of 397 case records. Pennsylvania has the largest percentage share of the entire CDR Survey, 20.9%, with over 2,000 case records being recorded.

THE SOCIO-ECONOMIC DATA

Once the actual waste variables from the survey had been chosen, an additional socio-economic data set was merged with the solid-waste information for analysis. The majority of the variables in this extra data set were chosen because they have been used in the solid-waste literature to help explain generation rates, or have been alluded to in the literature as having some effect upon generation rates.

These socio-economic variables were all taken from the 1972 County and City Data Book.[1] The variables chosen were:

> Median Family Income (MFI)
>
> Percentage of Income Less than $5,000 (LESS 5)
>
> Percentage of Income Greater than $15,000 (GREATER 1)
>
> Number of Families (FAMILIES)
>
> Number of Housing Units (HOUSING)
>
> Number of Persons per unit (PERSONUN)
>
> Density (DENSITY)

The following brief description of these variables indicates how they were determined and for what year they are applicable:

---

[1] County and City Data Book 1972: A Statistical Abstract Supplement, (Washington: U.S.G.P.O., U. S. Bureau of the Census, March 1973).

TABLE 11

NUMBER OF SAMPLE CASES BY STATE

State	Number of Cases	Percent of Sample
Alabama	7	1.8
Arizona	0	-
Arkansas	5	1.3
California	14	3.5
Colorado	1	.3
Connecticut	6	1.6
Delaware	1	.3
District of Columbia	0	-
Florida	7	1.8
Georgia	5	1.3
Idaho	1	.3
Illinois	1	.3
Indiana	4	1.0
Iowa	4	1.0
Kansas	6	1.6
Kentucky	3	.8
Louisiana	8	2.0
Maine	2	.5
Maryland	2	.5
Massachusetts	11	2.8
Michigan	16	4.0
Minnesota	1	.3
Mississippi	14	3.5
Missouri	16	4.0
Montana	3	.8
Nebraska	0	-
Nevada	1	.3
New Hampshire	0	-
New Jersey	19	4.8
New Mexico	3	.8
New York	18	4.5
North Carolina	9	2.3
North Dakota	9	2.3
Ohio	31	7.8
Oklahoma	6	1.6
Oregon	2	.5
Pennsylvania	84	20.9
Rhode Island	1	.3
South Carolina	9	2.3
South Dakota	8	2.0
Tennessee	12	3.0
Texas	23	5.8
Utah	2	.5
Vermont	0	-
Virginia	3	.8
Washington	8	2.0
West Virginia	7	1.8
Wisconsin	0	-
Wyoming	2	.5
Alaska	0	-
Hawaii	2	.5
	397	100.60*

* excess due to rounding

Median Family Income: Median total of all families income in dollars for the year 1969.[1]

Percentage of Income Less than $5,000: A sum total of the percentages of the families whose incomes were less than $3,000 and between the range of $3,000-$4,999, for the year of 1969.[2]

Percentage of Incomes Greater than $15,000: A sum total of the percentages of the families whose incomes were within the range of $15,000-$24,999 and above $25,000 for the year 1969.[3]

Number of Families: A family is defined as a household head and one or more other persons living in the same household who are related to the head by blood, marriage, or adoption; all persons in a household who are related to the head are regarded as members of his/her family.[4] Data for 1970.

Number of Occupied Housing Units: A housing unit is defined as occupied if it is the usual place of residence of the person or group of persons living within or at the time of census enumeration or if the occupants are only temporarily absent - for example, on vacation.[5] Data for 1970.

---

[1] Ibid.

[2] County and City Data Book, 1972: A Statistical Abstract Supplement, (Washington: U.S.G.P.O., U. S. Bureau of the Census). March, 1973.

[3] Ibid.

[4] Ibid.

[5] Ibid.

Density: Population per square mile is the average
number of inhabitants per square mile of land area.[1]
Data for 1970.

SOLID-WASTE GENERATION RATES: THEIR DEVELOPMENT

Once the waste and socio-economic variables had been
included in a common data set, attention was directed to deri-
vation of generation rates. It was not the intention of this
research to actually devise a new methodology for determining
solid-waste generation rates. On the contrary, it was to present new generation rates as determined from the 1968 National
Survey. Thus, the method used to obtain these generation rates
is the one described earlier in Chapter II, the population ratio
method. Basically, the method involves the division of the
total waste produced by residents of the area by the total
population of the area. The end result is a figure which sum-
marizes how much waste is produced per capita.

The solid-waste literature commonly reports generation
rates in pounds per person per day. In order to arrive at such
figures several conversions of the data had to be undertaken so
that consistent unit measurements of the data were available.
The most significant of these conversions was changing the
waste data measurements of tons and cubic yards to pounds.
This was accomplished by multiplying the waste tonnage figures
by the total number of pounds in a ton. The U. S. short ton
weight of 2,000 pounds avoirdupois was used. The problem of
converting cubic yards of waste into a poundage figure caused
additional problems. Depending on the amount of compaction and

---

[1]Ibid.

the actual constituents of the cubic yard of waste (i.e., steel, leaves, glass, paper, etc.) the weight of a cubic yard of waste varies substantially. Table 12 indicates how compaction (i.e. density) affects the weight per cubic yard of waste from twenty different geographical locations. Because of these differences a representative conversion factor had to be found.

Table 12 also indicates that the density of combined waste varied from a low weight of 300 pounds to a high of 750 pounds per cubic yard. The median was about 475 pounds per cubic yard, and this was the conversion factor used. Table 13 indicates that this figure of 475 was representative of eleven cities in the year 1965. At the same time, however, the table also indicates how time has affected the weight and density of solid wastes. With reference to this density problem the American Public Works Association (APWA) states:

> Rubbish density in compactor trucks varied from 350 to 700 pounds per cubic yard with a median of 590 pounds per cubic yard. The densities reported at Cincinnati indicate that noncombustible rubbish is more dense than combustible rubbish by about 100 pounds per cubic yard. The density of rubbish in open trucks was reported to be 222 pounds per cubic yard at Washington.

> Three cities reported the density of garbage collected in non-compactor trucks to range from 900 to 1,325 pounds per cubic yard. Philadelphia reported an average density of bulky wastes in open trucks of 200 pounds per cubic yard and of ashes in non-compactor trucks of 800 pounds per cubic yard.

The change in weight factors over time can be partially explained by the changing methods employed in packaging and the increased use of plastics. The APWA again states:

---

[1] Refuse Collections Practices, 3rd ed., (Chicago: American Public Works Association, 1966), p. 34.

TABLE 12

SOLID WASTE DENSITY, 1965

Class of Refuse	City	Density Lb/cu yd	Remarks[1]
Combined	Evanston, Ill.	525	
	Los Angeles, Calif.	500,570	Different size trucks
	Neptune, N.J.	600	
	Newark, N.J.	450	
	New York, N.Y.	390	Municipal collection
		480	Private collection
	Philadelphia, Pa.	450	
	San Francisco, Calif.	324	Estimated
	Seattle, Wash.	350,470	Different contractors
	Trenton, N.J.	450	
	Vancouver, B.C.	432	Different truck models
		582,683,712	
	Winnetka, Ill.	300	Commercial
		750	Residential
Rubbish	Atlanta, Ga.	400	
	Cincinnati, O.	580,600	Combustibles; different size trucks
		670,700	Noncombustibles; different size trucks
	Philadelphia, Pa.	350,509,625	Different size trucks
	Washington, D.C.	222	Open truck
Garbage	Los Angeles, Calif.	1,325	Non-compactor; restaurants only
	Philadelphia, Pa.	900	Non-compactor; residential only
	Washington, D.C.	1,000	Non-compactor
Bulky Wastes	Philadelphia, Pa.	200	Open trucks
Ashes	Philadelphia, Pa.	800	Non-compactor

[1] All densities were measured in compactor trucks unless otherwise noted.

Source: Refuse Collection Practices, 3rd Ed., (Chicago: American Public Works Association, 1966), p. 34.

TABLE 13

DENSITY OF VARIOUS SOLID WASTE CLASSIFICATIONS

(lb/cu yd)

	1939				1955				1965[1]			
	Cities	Min	Md	Max	Cities	Min	Md	Max	Cities	Min	Md	Max
Garbage	7	798	936	1540	-	-	-	-	3	900	1000	1325
Rubbish	5	200	214	677	5	60	383	605	3	350	590	700
Ashes	-	1150	1250	1400	-	-	-	-	1	-	800	-
Garbage and Combustible Rubbish	5	214	442	800	3	234	392	502				
Rubbish and Ashes	7	400	692	1000	-	-	-	-	1	580	-	600
Noncombustible Rubbish[2]	4	747	883	1165	3	396	543	605	1	670	-	700
Combined Refuse	6	500	604	1000	7	300	356	667	11	300	475	750

[1] In 1965 all densities are reported for compactor trucks except for garbage and ashes.

[2] Includes some ashes.

Data are from APWA surveys.

Source: Refuse Collection Practices, 3rd ed., (Chicago: American Public Works Association, 1966), p. 36.

Combined refuse showed a decrease in unit weight of
41 percent from 1939 to 1955 when the median value
declined from 604 to 356 pounds per cubic yard. The
causes are believed to be reduction in the operation
of both ashes and garbage and an increase in pro-
portion of combustible rubbish. This downward trend
in density has been significantly reversed as the
median value has now increased to 475 pounds per
cubic yard. Again, compaction appears to play the
predominant role.[1]

Thus, the development of generation rates from the

quantitative values of the waste data from the CRD, while still

using a traditional methodology, posed some problems. These

were rectified by using a tonnage conversion factor of 2000,

and a cubic yard conversion factor of 475, plus adjusting

the waste data to a daily quantity by using a conversion factor

of 365. It then was possible to use the traditional population

ratio method to obtain the desired generation rates in units of

pounds per person per day.

COMPUTER PROGRAMS USED IN ANALYSIS

A variety of computer programs was used in the research.

These were used to obtain access to the data, obtain sequential

file dumps, edit and move the data, develop the genera-

tion rates, obtain the sample, and create the socio-economic

data set. What follows is a brief description of these computer

programs:

PL/1. PL/1 was used to obtain access to the original
computer tapes in Cincinnati. It was also used to
duplicate the tapes.

COBOL. COBOL was used to 'unpack' the hexadecimal
format to a regular decimal format.

---

[1]Ibid., p.35.

SELECT.[1] SELECT was used to select the sample, and to create the sequential file dumps of the CDR and LDR.

SPSS. (Statistical Package for the Social Sciences).[2] SPSS was used to develop all of the generation rates and to perform the analyses of the CDR, the sample, and correlation and regression analyses. The following briefly describes the subprograms of SPSS which were used in the analysis of both the entire CDR and the sample, and in the correlation or regression analyses.

Frequencies: This was the first subprogram used in every case, providing the following: absolute frequency, relative frequency, adjusted frequency, and cumulative adjusted frequency. Likewise it provided all of the standard summary statistics such as standard deviation, mean, range, variance, number of observations, etc. to name a few.

Histograms: This provided a graphic plot of the value as provided.

Condescriptive: This subprogram provided only summary statistics.

Pearson Correlation: This subprogram computes the Pearson product-moment correlation for pairs of variables. For each of the variable pairs, a coefficient, number of cases, and significant levels are printed. A summary table for the means and standard deviations for each variable was also printed.

---

[1]Rich Jarvis, SELECT: A Utility Program to Manipulate a Sequential File, (Chicago: National Opinion Research Center, University of Chicago, September, 1972).

[2]Norman H. Nie, C. Hadlai Hull, et.al., Statistical Package for the Social Sciences(SPSS), Second Edition, (New York: McGraw-Hill Book Co., Inc., 1970).

Scattergrams: Subprogram scattergram will produce for each pair of variables a two-dimensional
plot (graph) of the data points where the coor-
dinates of the points are the values of the two
variables being considered.

Multiple Regression: To quote from the SPSS
manual, "Multiple regression is a general
statistical technique through which one can
analyze the relationship between a dependent
variable and a set of independent or predictor
variables. Multiple regression may be viewed
either as a descriptive tool by which the linear
dependence of one variable on the others is
summarized and decomposed, or as an inferential
tool by which the relationships in the population
are evaluated for the examination of the sample
data."[1]

Multiple regression analysis was performed to link the sample
waste data (dependent variables) with the socio-economic
variables (independent variables).

---

[1]Nie, Op. Cit., p.1.

CHAPTER IV

DATA ANALYSIS

Some cautionary remarks are in order before the computed generation rates are presented, because as noted earlier, the National Survey was far from a perfect data source, even if the only one available. Therefore, the first section of this chapter will deal with questions of the data quality which emerged during the data processing. Thereafter, we turn to the computed generation rates and to the factors associated with their variations.

DATA CHARACTERISTICS

Not all of the 12,142 case records of the CDR contained waste data for all of the categories. Although locational codes and population figures were present for each case, waste data were frequently absent for one or more waste categories.

Two population figures were indicated for each case record, a 1960 population figure and an estimated 1968 population figure. For the development of the generation rates it was decided to use the estimated 1968 population figure throughout, as the survey was taken during this year.

What then were the characteristics of the populations surveyed? EPA publications indicate that the National Survey was 75% urban.[1] This is a little misleading however, since

_____

[1] Anton J. Muhich, "Sample Representativeness and Community Data", The 1968 National Survey of Community Solid Waste Practices: An Interim Report (Washington: U. S. Environmental Protection Agency 1970).

these 'urban' populations included but a few large populated
areas.   Table 14 indicates the distribution of the population
by size class.   It shows that 59.2% of the areas surveyed were
under 5,000 people.   Similarily, 87.0% or 10,568 case records
out of a possible 12,142, fell into a population category of
20,000 people or less.   Only thirty cases or 0.2% of the survey,
were taken from areas whose populations were in the range of
500,000 to 1,000,000.   Clearly, the larger populated areas of
the country are not represented in the data.

Figure 7 indicates that the population category which
had the largest representation in the survey was in the population range of 1,001 to 5,000.   This represented 34.7% of the
total survey, or 4,217 cases.   The next largest population size
class was under 1,000 people, which comprised 24.4% of the
survey or 2,966 cases.   Collectively the population size classes
up to 5,000 persons accounted for 59.2% of the entire survey.
Clearly this skewed distribution of the population sampled
makes it very difficult to speculate about generation rates in
urban areas larger than 20,000 population.

Summary statistics for both the CDR and the sample indicated medians quite different from the mean, a large standard
error for all waste source categories, standard deviations in
the six figures, and data ranges from '0' to billions of pounds,
all indications of highly-skewed information.

TABLE 14

FREQUENCIES OF ESTIMATED 1968 POPULATION

POPULATION RANGES	VALUE	ABSOLUTE FREQUENCY	RELATIVE FREQUENCY (PERCENT)	ADJUSTED FREQUENCY (PERCENT)	CUMULATIVE ADJ. FREQ. (PERCENT)
Under 1000	1.00	2966	24.4	24.4	24.4
1001 - 5000	2.00	4217	34.7	34.7	59.2
5001 - 10000	3.00	1861	15.3	15.3	74.5
10001 - 20000	4.00	1525	12.6	12.6	87.0
20001 - 50000	5.00	1046	8.6	8.6	95.7
50001 - 100000	6.00	310	2.6	2.6	98.2
100001 - 500000	7.00	187	1.5	1.5	99.8
500001 - 1000000	8.00	30	0.2	0.2	100.0
	0.0	0	0.0	0.0	100.0
TOTAL		12142	100.0	100.0	100.0

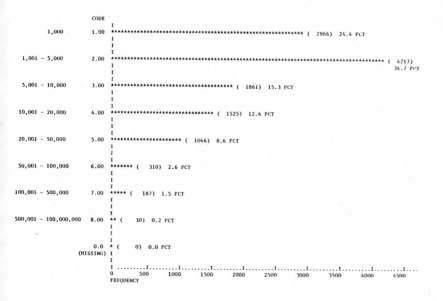

FIGURE 7

HISTOGRAM OF SURVEYED 1968 POPULATION

DERIVED SOLID-WASTE GENERATION RATES

Let us now recall the elements entering into the analysis. Waste measurements or estimations were derived from the National Survey of Community Solid Waste Practices. Population figures were for the year 1968 and were estimated as recorded on the survey forms. After converting the waste figures to pounds per day, the following national average generation rates were computed:

1. Household wastes (measured) 3.35 lbs/person/day

2. Commercial wastes (measured) 1.08 lbs/person/day

3. Combined wastes (measured) 3.48 lbs/person/day

Only the measured waste values were used as there were too few cases of the estimated.

These national average generation rates were determined by using the population ratio method with adjustments for measurement conversions and the number of days in the year.

Closer inspection of the generation rates indicated that there were often several extreme rates which were included in the averaging procedure. These rates were eliminated from the averaging procedure as none of the literature reviewed in the remotest sense indicated per capita generation rates in these ranges. Examples of these extreme generation rates were: 34.70 lbs/person/day for combined, and 15.35 lbs/person/day for household. Thus, it was decided to exclude from further data analysis any generation rate of 10.00 lbs/person/day or over, as most likely indicative of poor basic data.

Taking the derived national generation rates and eliminating these extreme values provided adjusted generation rates.

These rates are:

1. Household wastes   2.49 lbs/person/day

2. Combined wastes   2.94 lbs/person/day

STUDY FIGURES/U.S. EPA FIGURES

These generation rates were then compared to the generation rates published by the EPA. These comparisons, as seen in Table 15, indicate that EPA's published figures for urban, rural, and national categories are all well below the derived figures from the sample. This is true for all their waste source categories of household, commercial, and combined. Even sub-totaling the waste source categories household and commercial produced sample figures consistently higher than the EPA's.

Additional generation rate computations were made by individual states (see Table 16). The sample-derived generation rates were plotted by individual case record on state outline maps of the United States. As before, extreme generation rates (i.e., over 10.0 lbs/person/day) were eliminated.

REGIONAL VARIATION

It was thought that these generation-rate values would indicate some regional variation in solid-waste generation as some literature suggests. The results however, were inconclusive. No regional variation of waste generation was indicated.

TABLE 15

GENERATION-RATE COMPARISONS

lbs/person/day

Waste Source	Urban	Rural	National	Sample	%Increase over EPA Urban	%Increase over EPA Rural	%Increase over EPA National
Household* (measured)	1.26	0.72	1.14	$\frac{3.35}{2.49}$	$\frac{166}{73}$	$\frac{365}{204}$	$\frac{194}{92}$
Commercial (measured)	0.46	0.11	0.38	1.08	134	888	184
Combined* (measured)	2.63	2.60	2.63	$\frac{3.48}{2.94}$	$\frac{32}{12}$	$\frac{32}{13}$	$\frac{32}{12}$
Sub Totals							
Household	1.26	0.72	1.14	3.35/2.19			
Commercial	0.46	0.11	0.38	1.08/1.08			
	1.72	0.83	1.52	4.43/3.27			

*The dual figures indicate sample-derived values before and after extreme values
were eliminated from averaging procedure. Bottom figure($\frac{3.35}{2.49}$) is the adjusted figure.
(e.g.

STATE GENERATION RATES

As Table 16 indicates in nearly all instances, the state averages for solid-waste generations were higher than the previously mentioned EPA figures. At the same time however, they were often lower than the sample average. The individual case record for any state indicates a wide range of generation-rate values for any or all of the waste source categories.

There were twenty-six (26) states averaged for household (measured), twelve (12) states for commercial (measured), and 32 states for combined (measured).

Why were the sample and many of the state generation rate averages in excess of the EPA figures? Were there faults in the methods used in determining the sample-derived generation rates? Were invalid conversion factors used? Were the methods used by the EPA subject to error?

DENSITY CONVERSION FACTOR CHECK

A review of the methodologies used by both the EPA and this research indicated that a weak area might be in the value used as the density conversion factor. If the conversion factors initially used by this research were too high, then the obtained generation rates would possibly also be higher than the EPA rates. These higher rates would especially be true if EPA's density conversion factors were less than this research. Thus, it was necessary to test what the effects would be of a reduced density conversion factor on generation rates.

The density conversion factor was reduced from 475 lbs/ cubic yard to 275 lbs/cubic yard of waste, a reduction of 200 pounds. From available sources, this was the lowest figure which could be used as representing compacted waste.

TABLE 16

GENERATION RATES AND NUMBER OF CASES BY STATE OF THE SAMPLE

State	Household		Commercial		Combined	
	GR	# Cases	GR	# Cases	GR	# Cases
Alabama	$\frac{5.72}{1.24}$	$\frac{2}{1}$				
Alaska						
Arizona	4.04	2				
Arkansas			7.90	2		
California					$\frac{8.02}{3.50}$	$\frac{4}{3}$
Colorado					8.58	
Connecticut	3.25	3			3.37	2
Delaware	2.64	2			4.69	17
Florida	1.49	1			2.40	5
Georgia	$\frac{8.26}{2.20}$	$\frac{2}{1}$	2.13	2	$\frac{5.14}{2.63}$	$\frac{3}{2}$
Hawaii	2.40	2	1.52	2		
Idaho						
Illinois					1.47	2
Indiana	8.16	1				
Iowa	0.64	2			0.48	1
Kansas					$\frac{11.28}{2.95}$	$\frac{6}{5}$
Kentucky					2/09	2
Louisiana	1.95	1	1.64	2	1.96	8
Maine					3.44	2
Maryland					6.72	
Massachusetts	4.08	3	0.99	2	$\frac{7.08}{3.12}$	$\frac{3}{2}$
Michigan			0.88	4	3.85	7
Minnesota					5.87	1
Mississippi	2.01	11	0.94	14	.62	2
Missouri						
Montana						

TABLE 16
(cont'd)

GENERATION RATES AND NUMBER OF CASES BY STATE OF THE SAMPLE

State	Household		Commercial		Combined	
	GR	# Cases	GR	# Cases	GR	# Cases
Nebraska						
New Hampshire						
New Jersey						
New Mexico					6.51	1
New York	3.24	6	1.02	4	5.54	5
N. Carolina	4.76	2	0.92	2	3.28	9
N. Dakota	1.63	1			5.87	5
Ohio	1.93	1			2.95	27
Oklahoma					2.63	5
Oregon	2.23	1			1.92	2
Pennsylvania	3.36 / 1.82	31 / 29	0.36	25	1.99	47
Rhode Island	1.83	1				
S. Carolina					1.72	6
S. Dakota	1.69	5	1.80	5	1.37	1
Tennessee	6.86 / 2.62	3 / 2	4.45	2	14.12 / 8.42	2 / 1
Texas	1.84	2			6.98	19
Utah						
Vermont						
Virginia	8.15 / 4.00	2 / 1				
Washington					9.27 / 4.64	6 / 5
W. Virginia					4.09	7
Wisconsin						
Wyoming	13.01	1				

Table 17 indicates the comparison of generation rates, values as derived from using both density conversion factors. Note that there was no substantial drop in generation-rate value and that the values derived from using the 275 lbs/cubic yard conversion factor were still consistently higher than the EPA figures. It was concluded therefore that the conversion factors used in the research did not significantly influence the differences in values between the sample-derived rates and the EPA rates.

TABLE 17

DENSITY CONVERSION FACTOR EFFECT ON GENERATION RATES

Waste Source	Generation Rates		
	Research Sample		EPA
	275 Density lb/cubic yd	475 Density lb/cubic yd	National EPA Figures
Household (measured)	2.34 lb/ cap./day	2.49 lb/ cap./day	1.15 lb/ cap./day
Commercial (measured)	0.84 lb/ cap./day	1.08 lb/ cap./day	0.38 lb/ cap./day
Combined (measured)	2.89 lb/ cap./day	2.94 lb cap./day	2.63 lb cap./day

GENERATION-RATE CHECK WITH 31 VIRGINIA CITIES

To further test the sample results, generation-rates were compared with similarly derived rates from 31 of the 34 independent cities of Virginia (See Table 18). The results indicated that the sample-derived national generation rates were very low when compared to the Virginia cities.

For the waste-source category of combined (measured),
and with 18 of the 31 Virginia cities having data, the average
generation rate was 6.50 lbs/person/day. This compared to the
sample average of 2.94 lbs/person/day, and an EPA National
Average of 2.63 lbs/person/day. For household (measured), and
with 10 out of the 31 Virginia cities having data, a generation
rate of 7.00 lbs/person/day was obtained. Compare this figure
with the sample-derived rate of 2.49 lbs/person/day, and the
EPA National average of 1.14 lbs/person/day. Thus, even the
sample-derived rates may be low relative to specific local
conditions.

TABLE 18

INDEPENDENT CITIES OF VIRGINIA
(-) not included

1. Alexandria	18. Lynchburg
2. Bristol	19. Martinsville
3. Buena Vista	20. Newport News
4. Charlottesville	21. Norfolk
5. Chesapeake	22. Norton (-)
6. Clifton Forge	23. Petersburg
7. Colonial Heights	24. Portsmouth
8. Covington	25. Radford
9. Danville	26. Richmond
10. Fairfax	27. Roanoke
11. Falls Church	28. South Boston (-)
12. Franklin	29. Staunton
13. Fredericksburg	30. Suffolk (-)
14. Galax	31. Virginia Beach
15. Hampton	32. Waynesboro
16. Harrisonburg	33. Williamsburg
17. Hopewell	34. Winchester

SOCIO-ECONOMIC AND SOLID-WASTE GENERATION-RATE CORRELATES

How do socio-economic factors correlate with solid-waste generation rates? The evidence appears in Table 19. Each of the
waste-source categories of the sample -- household (estimated
and measured), commercial (estimated and measured), and combined (estimated and measured) -- was correlated with several
selected socio-economic variables -- median family income (MFI),
percentage of incomes less than $5,000 (LESS), percentage of in-
comes greater than $15,000 (GREATER 1), number of families
(FAMILIES), number of housing units (HOUSING), number of persons per unit (PERSONUN), density (DENSITY), the 1960 population (POP 60), and the estimated 1968 population (POP 68).
These social factors were chosen because they were the ones
most often referred to in the literature as variables influencing solid-waste generation.

The correlation coefficients between waste-source variables and the socio-economic variables vary widely. The figures
are shown in Table 19. The strongest correlations are between
social variables and the measured waste-source categories, not
the estimated. The socio-economic variables which had the largest number of correlations statistically significant at the 95%
confidence level were HOUSING, FAMILIES, DENSITY, and PERSONUN.
Most of the inverse relationships occur in the "estimated"
positions of the household and commercial waste-source categories.

MULTIPLE LINEAR REGRESSION MODELS

How do the statistically significant social variables
influence the generation rates? This question was explored using

TABLE 19

CORRELATION COEFFICIENTS

Waste Source Category	Pop 68	Pop 60	Density	Personun	MFI	Less 5	Greater 1	Families	Housing
Household Waste Generation Rates (Estimated)	0.0405 (13)	0.0611 (13)	-0.4032 (13)	0.0474 (13)	-0.5053* (13)	0.5001* (13)	-0.4907* (13)	-0.2865 (13)	-0.2816 (13)
Household Waste Generation Rates (Measured)	0.5685* (115)	0.5485* (115)	0.0680 (115)	0.1699* (115)	0.1649* (115)	-0.0864 (115)	0.1440 (115)	0.0531 (115)	0.0342 (115)
Commercial Waste Generation Rates (Estimated)	0.2023 (15)	-01943 (15)	-0.1878 (15)	0.2491 (15)	-0.4716* (15)	0.5467* (15)	0.4641* (15)	-0.1357 (15)	-0.1333 (15)
Commercial Waste Generation Rates (Measured)	0.7206* (95)	0.6622* (95)	0.0497 (95)	0.2837* (95)	0.1773* (95)	-0.0916 (95)	0.2002* (95)	0.1008 (95)	0.0759 (95)
Combined Waste Generation Rates (Estimated)	0.7899* (24)	0.8095* (24)	0.5770* (24)	-0.2845* (24)	0.4230* (24)	-0.4073* (24)	0.4064* (24)	0.2210 (24)	0.1973 (24)
Combined Waste Generation Rates (Measured)	0.8292* (235)	0.8072* (235)	0.2218* (235)	0.1338* (235)	0.2564* (235)	-0.1836 (235)	0.2927* (235)	0.0523 (235)	0.0417 (235)

(0) Number of Cases
 * Significant at the .05 Level

multiple linear regression models.

The regression models used each of the waste-source categories as the dependent variables. The following socio-economic variables. were initially considered as the independent variables in the preliminary runs: MFI (median family income), LESS 5 (percent of incomes greater than $15,000), FAMILIES (Number of families), HOUSING (the number of housing units), PERSONUN (the number of persons per unit), DENSITY (density), POP 68 (the 1968 population). However, scattergrams indicated that a great deal of multicollinearity existed between several of the independent variables, thus the linear regressions models were finally developed using only MFI, DENSITY, and POP 68 as the independent variables. The results of these regression models are summarized in Table 20.

This table indicates, for each of the independent vari-ables, the regression coefficients of the linear model, the standard errors of the linear regression coefficients, and the regression coefficients of the log-transformed regression model. Natural log transformations of all variables in the models were performed to attempt to account for any bivariate or multi-variate skews in the latter case. The effects of these log transformations on the partial regression coefficients can be seen in Table 20.

The results of the linear regression models are as follows: The strongest association occurred in the regression model which had as a dependent variable the waste-source category of combined (measured) and the independent variables of MFI, DENSITY, and POP 68. The model had an R^2 of 0.693. The waste-source category of combined (estimated) and its in-dependent variables of MFI, DENSITY and POP 68, was the second

TABLE 20

PARTIAL REGRESSION COEFFICIENTS

Dependent Variables	Independent Variables			R²
	Population 1968	Density	Median Family Income	
Household (Estimated)	0.00001(1) 0.00001(2) (-0.02985)(3)	-0.00014 0.00093 (0.26232)	-0.00058 0.00038 (-14.43827)	0.34892 ------ 0.28665
Household (Measured)	0.36324 0.05184 (0.46622)	0.60071 1.57054 (0.88654)	0.48870 1.01421 (-3.94761)	0.32614 ------ 0.14260
Commercial (Estimated)	0.00000 0.00007 (-0.22198)	0.00613 0.00556 (0.37200)	-0.00428 0.00218 (-16.40121)	0.30025 ------ 0.43824
Commercial (Measured)	0.31198 0.03245 (0.68184)	0.09075 0.97036 (0.61834)	-0.14380 0.64731 (-5.07407)	0.51952 ------ 0.14289
Combined (Estimated)	0.53424 0.11650 (0.21366)	7.64182 4.18295 (1.63400)	-3.61445 3.51800 (-11.93471)	0.67839 ------ 0.22519
Combined (Measured)	0.79655 0.03687 (0.75263)	2.48507 1.82359 (0.43134)	0.28667 0.63636 (-3.04796)	0.69306 ------ 0.15470

(1) Regression coefficient in the LINEAR model
(2) STANDARD ERROR of the LINEAR Regression COEFFICIENT
(3) Regression coefficient in the LOG TRANSFORMED

most powerful, with a regression model R^2 of 0.678. The other regression models had R^2 which ranged from 0.519 to 0.348. The log-transformed models generally had much lower R^2. All models indicate that the greater solid-waste generation rates occur in areas with large populations, with high densities, and occupied by lower-income groups. Likewise, lower solid-waste generation rates occur in small, low density, high-income communities.

Some of the generation-rate relationships are depicted in Figure 8. In this figure, each of the measured waste-source categories equations has been plotted and graphs have been constructed showing the partial relationships in these equations between the dependent and each of the independent variables, controlling for the others.

The inverse relationship between solid-waste generation and median family income is apparent. This inverse relationship is present in nearly all the regression models using MFI as a dependent variable (see Table 20 and Figure 8).

Likewise, the positive relationship between waste generation and both population and population density also is apparent.

Clearly, these relationships do not reflect the total significance of the regressions results. If one solves the regression equation, then each of the regression coefficients of the log transformations can be viewed as a percentage increase or decrease due to the one independent variable while at same time controlling for the other two variables. For example, if we write a natural log-transformed equation for household (measured) it reads:

FIGURE 8

GRAPHIC TRENDS BETWEEN INDEPENDENT AND DEPENDENT VARIABLES*

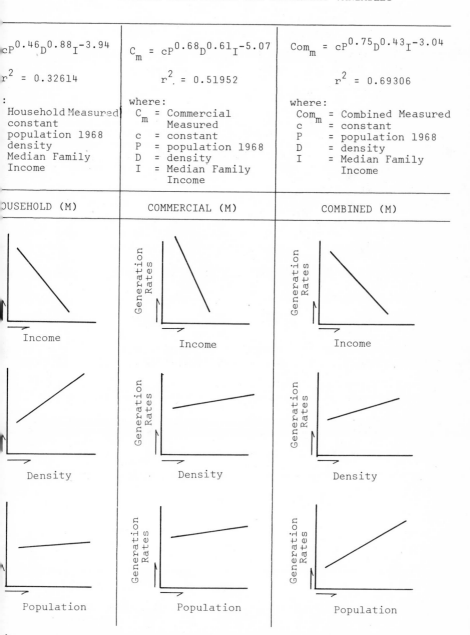

$cP^{0.46}D^{0.88}I^{-3.94}$

$r^2 = 0.32614$

:
Household Measured
constant
population 1968
density
Median Family
Income

$C_m = cP^{0.68}D^{0.61}I^{-5.07}$

$r^2 = 0.51952$

where:
C_m = Commercial
        Measured
c = constant
P = population 1968
D = density
I = Median Family
    Income

$Com_m = cP^{0.75}D^{0.43}I^{-3.04}$

$r^2 = 0.69306$

where:
Com_m = Combined Measured
c = constant
P = population 1968
D = density
I = Median Family
    Income

OUSEHOLD (M)   COMMERCIAL (M)   COMBINED (M)

*ALL TREND LINES APPROXIMATED

$$\ln H_m = \ln c + 0.46 \ln P + 0.88 \ln D - 3.94 \ln I$$

where;

      $\ln$ = Natural Logarithm

      H_m = Household (measured)

      c = constant (34.02419) (34.02 rounded)

      P = population 1968

      D = density

      I = median family income

If we write the equation for Commercial (measured) it reads:

$$\ln C_m = \ln c + 0.68 \ln P + 0.61 \ln D - 5.07 \ln I$$

where;

      $\ln$ = Natural Logarithm

      C_m = Commercial (measured)

      c = constant (42.03290) (42.03 rounded)

      P = population 1968

      D = density

      I = median family income

Likewise, the equation for combined (measured) reads:

$$\ln COM_m = \ln c + 0.75 \ln P + 0.43 \ln D - 3.04 \ln I$$

where;

      $\ln$ = Natural Logarithm

      COM_m = Combined (measured)

      c = constant (26.87258) (26.87 rounded)

      P = population 1968

      D = density

      I = median family income

REGRESSION MODELS SOLVED FOR THREE SELECTED CITIES

These equations can be used as a predictive tool when applied to cities of different sizes each having different densities and median family incomes. Let us call a sample of cities "A", "B", "C".

CITY "A"

City "A" has a population of 5,000, a density of 10 persons per acre (6400 persons per square mile), and a median family income of $7,000. If we use these values to solve the previous equations, the following results are obtained:

The Household measured (H_m) equation is:

$\ln H_m = \ln c + 0.46 \ln P + 0.88 \ln D - 3.94 \ln I$

Substitutions provide the following:

$\ln H_m = 34.02 + 0.46(8.52) + 0.88(8.76) - 3.94(8.85)$

$\ln H_m = 34.02 + 3.92 + 7.74 - 34.87$

$\ln H_m = 45.68 - 34.87$

$\ln H_m = 10.81$

(antilog)$H_m = 49,513.47$ lbs/year

The Commercial measured (C_m) equation is:

$\ln C_m = \ln c + 0.68 \ln p + 0.61 \ln d - 5.07 \ln I$

Substitutions provide the following:

$\ln C_m = 42.03 + 0.68(8.52) + 0.61(8.76) - 5.07(8.85)$

$\ln C_m = 42.03 + 5.79 + 5.34 - 44.87$

$\ln C_m = 53.16 - 44.87$

$\ln C_m = 8.29$

(antilog) $C_m = 3983.83$ lbs/year

The Combined measured (COM_m) equation is:

$\ln COM_n = \ln c + 0.75 \ln p + 0.43 \ln D - 3.04 \ln I$

Substitutions provide the following:

$lnCOM_m = 26.87 + 0.75(8.52 + 0.43(8.76) - 3.04(8.85)$

$lnCOM_m = 26.87 + 6.39 + 3.77 - 26.90$

$lnCOM_m = 37.03 - 26.90$

$lnCOM_m = 10.13$

$(antilog)COM_m = 25,084.36$ lbs/year

CITY "B"

City "B" has a population of 25,000, a density of 20 persons per acre, (12,800 persons per square mile) and a median family income of $12,000. Substitution of these values into the household measured equations provides the following:

The Household measured (H_m) equation is:

$lnH_m = 34.02 + 0.46(10.13) + 0.88(9.46) - 3.94(9.39)$

$lnH_m = 34.02 + 4.66 + 8.32 - 36.99$

$lnH_m = 47.00 - 36.99$

$lnH_m = 10.01$

$(antilog)H_m = 22,247.84$ lbs/year

The Commercial measured (C_m) equation is:

$lnC_m = 42.03 + 0.68(10.13) + 0.61(9.46) - 5.07(9.39)$

$lnC_m = 42.03 + 6.89 + 5.77 - 47.60$

$lnC_m = 54.69 - 47.60$

$lnC_m = 7.09$

$(antilog)C_m = 1,199.91$ lbs/year

The Combined measured (COM_m) equation is:

$lnCOM_m = 26.87 + 0.75(10.13) + 0.43(9.46) - 3.04(9.39)$

$lnCOM_m = 26.87 + 7.60 + 4.06 - 28.55$

$lnCOM_m = 38.43 - 28.55$

$lnCOM_m = 9.88$

$(antilog)COM_m = 19,535.72$ lbs/year

CITY "C"

City "C" has a population of 50,000, a density of 50 persons per acre (32,000 persons per square mile) and a median family income of $15,000.

Substitution of these values provides the following:

The Household measured (H_m) equation is:

$lnH_m = 34.02 + 0.46(10.82) + 0.88(10.37) - 3.94(9.62)$

$lnH_m = 34.02 + 4.98 + 9.13 - 37.90$

$lnH_m = 48.13 - 37.90$

$lnH_m = 10.23$

$(antilog)H_m = 27,722.51$ lbs/year

The Commercial measured (C_m) equation is:

$lnC_m = 42.03 + 0.68(10.82) + 0.61(10.37) - 5.07(9.62)$

$lnC_m = 42.03 + 7.36 + 6.33 - 48.77$

$lnC_m = 55.72 - 48.77$

$lnC_m = 6.95$

$(antilog)C_m = 1,043.15$ lbs/year

The Combined measured (COM_m) equation is:

$lnCOM_m = 26.87 + 0.75(10.82) + 0.43(10.37) - 3.04(9.62)$

$lnCOM_m = 26.87 + 8.11 + 4.46 - 29.24$

$lnCOM_m = 39.44 - 29.24$

$lnCOM_m = 10.20$

$(antilog) COM_m = 26,903.19$

Each of the equations (i.e. Household measured (H_m), Commercial measured (C_m), and Combined measured (COM_m) was solved by doubling each of the independent variables separately while holding others constant.

The results follow:

City "A"

Characteristics

	Baseline Value	ln Natural Logarithm	Doubled Value	ln Natural Logarithm
Population =	5,000	8.52	10,000	9.21
Density =	10/acre or 6400/sq. mile	8.76	12,800	9.46
MFI	7,000	8.85	14,000	9.55

Baseline

lnH_m - 34.02 + 0.46(8.52) + 0.88(8.76) - 3.94(8.85)

lnH_m = 34.02 + 3.92 + 7.71 - 34.87

lnH_m = 45.65 - 34.87

lnH_m = 10.78

(antilog)H_m = 48,050.12

Population Doubles

lnH_m = 34.02 + 0.46(9.21) + 0.88(8.76) - 3.94(8.85)

lnH_m = 34.02 + 4.23 + 7.71 - 34.87

$lnHm$ = 45.96 - 34.87

lnH_m = 11.09

(antilog)H_m = 65,512.74

                                     73.3% increase over baseline

Density Doubles

lnH_m = 34.02 + 0.46(8.52) + 0.88(9.46) - 3.94(8.85)

lnH_m = 34.02 + 3.92 + 8.32 - 34.87

lnH_m = 46.26 - 34.87

lnH_m = 11.39

(antilog)H_m = 88,432.96

54.33% increase over baseline

MFI Doubles

$\ln H_m$ = 34.02 + 0.46(8.52) + 0.88(8.76) - 3.94(9.55)

$\ln H_m$ = 34.02 + 3.92 + 7.71 - 37.63

$\ln H_m$ = 45.65 - 37.63

$\ln H_m$ = 8.02

(antilog)H_m = 3,041.18

93.67% decrease over baseline

For City "A" and Household measured (H_m), the following results were obtained:

a. A doubling of population would produce a 73.30% increase in solid-waste generation rates, controlling for density and income.

b. A doubling of density would produce a 54.33% increase in solid-waste generation rates, controlling for population and income.

c. A doubling of the median family income would produce a 93.67% decrease in solid-waste generation rates, controlling for population and density.

City "A" (Commercial$_m$)

Baseline   Constant 42.03

$\ln C_m$ = 42.03 + 0.68(8.52) + 0.61(8.76) - 5.07(8.85)

$\ln C_m$ = 42.03 + 5.79 + 5.34 - 44.87

$\ln C_m$ = 53.16 - 44.87

$\ln C_m$ = 8.29

(antilog)C_m = 3983.83

Population Double

$\ln CM = 42.03 + 0.68(9.21) + 0.61(8.76) - 5.07(8.85)$

$\ln_m = 42.03 + 6.26 + 5.34 - 44.87$

$\ln C_m = 53.62 - 44.87$

$\ln C_m = 8.75$

$(antilog) C_m = 6310.69$

> 63.13% increase over baseline

Density Doubles

$\ln C_m = 42.03 + 0.68(8.52) + 0.61(9.46) - 5.07(8.85)$

$\ln C_m = 42.03 + 5.79 + 5.77 - 44.87$

$\ln C_m = 53.59 - 44.87$

$\ln C_m = 8.72$

$(antilog\ C_m = 6124.18$

> 65.605% increase over baseline

MFI Doubles

$\ln C_m = 42.03 + 0.68(8.52) + 0.61(8.76) - 5.07(9.55)$

$\ln C_m = 42.03 + 5.79 + 5.34 - 48.42$

$\ln C_m = 53.16 - 48.42$

$\ln C_m = 4.74$

$(antilog) C_m = 114.43$

> 97.13% decrease over baseline

The results obtained for City "A" with the Commercial measured (C_m) model were:

a. A doubling of population would produce a 63.13% increase in solid-waste generation rates, controlling for density and income.

b. A doubling of density would produce a 65.05% increase in solid-waste generation rates, controlling for population and income.

c.   A doubling of the median family income would produce a 97.13%
<u>decrease</u> in solid-waste generation rates, controlling for population and
density.

City "A" (combined$_m$)

<u>Baseline</u>

$\ln COM_m = 26.87 + 0.75(8.52) + 0.43(8.76) - 3.04(8.85)$

$\ln COM_m = 26.87 + 6.39 + 3.77 - 26.90$

$\ln COM_m = 37.03 - 26.90$

$\ln COM_m = 10.13$

$(antilog) COM_m = 25,084.36$

<u>Population Doubles</u>

$\ln COM_m = 26.87 + 0.75(9.21) + 0.43(8.76) - 3.04(8.85)$

$\ln COM_m = 26.87 + 6.91 + 3.77 - 26.90$

$\ln COM_m = 37.55 - 26.90$

$\ln COM_m = 10.65$

$(antilog) COM_m = 42,192.59$

$\hspace{6cm}$ 59.45% increase over baseline

<u>Density Doubles</u>

$\ln COM_m = 26.87 + 0.75(8.52) + 0.43(9.46) - 3.04(8.85)$

$\ln COM_m = 26.87 + 6.39 + 4.07 - 26.90$

$\ln COM_m = 37.33 - 26.90$

$\ln COM_m = 10.43$

$(antilog) COM_m = 33,860.35$

$\hspace{6cm}$ 74.08% increase over baseline

<u>MFI Doubles</u>

$\ln COM_m = 26.87 + 0.75(8.52) + 0.43(8.76) - 3.04(9.55)$

$\ln COM_m = 26.87 + 6.38 + 3.77 - 29.03$

$\ln COM_m = 37.02 - 29.03$

$\ln COM_m = 7.99$

$(antilog) COM_m = 2951.30$

88.23% decrease over baseline

The Combined measured (COM_m) model for City "A" indicated the following:

a. A doubling of population would produce a 59.45% increase in solid-waste generation rates, controlling for density and income.

b. A doubling of density would produce a 74.08% increase in solid-waste generation rates, controlling for population and income.

c. A doubling of median family income would produce a 88.23% decrease in solid-waste generation rates, controlling for population and density.

Thus, for each of the City "A" models, the greatest positive effect was determined by the doubling of the density variable. The next greatest positive effect was caused by the doubling of the population variable. In each model however, a doubling of the median family income had the effect of dramatically decreasing solid-waste generation.

If we take the results of the natural log predictive models for all three cities, "A", "B", "C", and use those results as a representative value of the total amount of solid waste generated for those cities, we can arrive at a daily generation rate by simple division. The results of this are shown in Table 21.

In comparison, the predictive rates for city "A" are generally higher than similar rates as determined earlier in this research (Table 17) and/or National E.P.A. figures. On the other hand, cities "B" and "C" were lower than those figures

TABLE 21

PREDICTED GENERATION RATES FOR

THREE CITIES

City "A"

H_m = 49,513.47 ÷ 5,000 = H_m Generation Rate = 9.90 lbs/
        capita/day

C_m = 3,983.83 ÷ 5,000 = C_m Generation Rate = 0.78 lbs/
        capita/day

COM_m = 25,084.36 ÷ 5,000 = COM_m Generation Rate = 5.02 lbs/
        capita/day

City "B"

H_m = 22,247.84 ÷ 25,000 = H_m Generation Rate = 0.88 lbs/
        capita/day

C_m = 1,199.91 ÷ 25,000 = C_m Generation Rate = 0.04 lbs/
        capita/day

COM_m = 19,535.72 ÷ 25,000 = COM_m Generation Rate = 0.78 lbs/
        capita/day

City "C"

H_m = 27,722.51 ÷ 50,000 = H_m Generation Rate = 0.55 lbs/
        capita/day

C_m = 1,043.15 ÷ 50,000 = C_m Generation Rate = 0.02 lbs/
        capita/day

COM_m = 26,903.19 ÷ 50,000 = COM_m Generation Rate = 0.53 lbs/
        capita/day

shown in Table 17.   Why were these results obtained?

It must be recalled that the results of these equations are based upon data from small communities whose populations are not large (see Figure 7) and whose population densities are presumably very low. In communities such as these the idea of a decrease in waste generation with higher incomes is probable, more so than in larger more dense communities whose populations are in the range of 400,000 plus. The homogeneity of populations in small communities therefore, probably plays a significant role in determining solid waste generation rates.   One might naturally suspect that increasing population and density would have positive affects on the amount of solid waste generated. One might not suspect however a decrease in waste generated with higher incomes.

CHAPTER V
CONCLUSIONS

Before discussing the specific conlusions of this research, I feel compelled to review the past and present feelings
that government and individuals have had towards solid waste.

While the old adage, "the illusion of diffusion" may
apply to water and air pollution, the adage, "out of sight, out
of mind" certainly applies to solid waste. People seem to be-
lieve this adage as they toss beer cans out of their car windows.
They only become concerned about solid waste when their weekly
'garbage' is not picked up. Cities it seems are only concerned about how to collect, transport, bury or mash solid waste.
This is clearly the opposite viewpoint to that taken by people who
want to stop water and air pollution. Their cry is clear:
stop pollution at its source! Why then has solid waste been
"managed"? Why haven't we been concerned about the source of
solid-waste generation? A partial answer to this might be obtained by reveiwing the Federal government's role in the solid
waste field.

Federal Government's Involvement in Solid Waste

The Federal Government's role in the solid-waste field
was not firmly established until after World War II. Before
that time private companies or individuals had done little
research into solid-waste generation. Much was written however, about what to do with the waste after it had been 'generated.' Even after World War II, only a very small staff of
less than six persons was responsible for the development of
guidelines on 'how to' bury, collect, and burn waste.

The Public Health Service, in cooperation with the American Public Works Association (APWA), sponsored a national conference on solid-waste research at Chicago, Illinois in 1963. Two manuals were the outcome of that conference; Municipal Refuse Disposal[1] and Refuse Collection Practices[2], both of which are still widely quoted.

President Johnson, in 1965, recommended legislation to sponsor demonstration projects and to establish more effective methods of waste disposal and recycling. The end result of the Congressional action on these proposals was the Solid Waste Disposal Act of 1965 (Public Law 89-272), and its ammended version in 1968.

Through time, the agencies responsible for the administration of these laws changed often, until 1970 when the Environmental Protection Agency was formed. As a result of these reorganizations and redirections of effort, many programs suffered; the 1968 National Survey of Community Solid Waste Practices was one of these. It has suffered to the extent today that the Solid Waste Branch of the EPA is small when compared to the other branches.

Thus, while the Federal Government's role was to 'accommodate' waste, solid-waste problems increased. Little or no concern was given to the rising amount of waste being generated. This was because solid waste was never really considered as a pollutant, as it was always 'managed' and accommodated.

---

[1]Municipal Refuse Disposal, 3rd edition, (Chicago: American Public Works Association, 1970).

[2]Refuse Collection Practices, 3rd edition (Chicago: American Public Works Association, 1966).

Since 'Earth Day' 1970, and the beginning of public
awareness to environmental issues, the interest in solid waste
has increased.

Garage sales have reduced some of the volume of solid
wastes with which cities have had to contend. Indeed, a reduction
of as much as 10% has been noticed by some communities. This is
not totally due to garage sales, but also to the state of the
economy. When more people have less to spend, they buy less
and throw less away.

Technology has helped to reshape the thinking about
solid waste. Cities are now beginning to view solid waste as
an asset rather than a liability. They can now sell their
waste to such companies as Consolidated Edison or other similar
electrical power generating facilities. Indeed these companies
now have the ability to separate, pulverize, dry, and burn
solid waste. This one-time nuisance of a pollutant is now a
valued fuel.

Helping this movement obviously has been the energy
crisis and government decrees to make America self-sufficient
in fuels by 1980. Thus, the economic incentive to use solid
waste has come to the fore. Likewise; the increased costs of
raw materials makes it very desirable to recycle solid wastes.

All of these and other factors have shed new light upon
the solid waste field. Communities are now very interested in
just how much waste they generate and how it can be turned into
a commodity which can easily be marketed. Indeed, if events
keep occuring at the rate that they have, industries may soon
be responsible for the collection and disposal of all municipal
wastes.

As the <u>1968 National Survey of Community Solid Waste Practices</u> is the only national solid-waste survey ever attempted, any data which can be resolved is worthwhile. Problems inherent with missing data, lack of uniformity, the inability to compare and contrast data from one questionnaire to another, and the small population data base from which the data was collected, do not negate the value of the available information. This problematic source simply indicates by its own existence that further research is needed.

The conclusions, however, must be viewed in perspective. They are conclusions from data nearly nine years old. As such, changes that have taken place in society and the economy which affect solid waste generation rates should be considered when trying to up-date these conclusions to the present.

In short therefore, this research has attempted to partially fill a data gap that is very apparent in the solid-waste field. This was done by analyzing a portion of the data of a "one of a kind" survey, the <u>1968 National Survey of Community Solid Waste Practices</u>, to arrive at what are felt to be more accurate and meaningful generation rates for the waste source categories of household, commercial and combined, and to explore the relationships which exist between certain socio-economic variables and solid-waste generation for the year 1968.

PURPOSE OF RESEARCH

As indicated in Chapter I, the purpose of this research was:

1. To analyze data collected in the 1968 National Survey of Community Solid Waste Practices.

2. To determine actual rather than estimated amounts of
solid wastes being generated.

3. To determine the variations existing between the
rates of solid-waste generation by source.

4. To determine what effects socio-economic variables
may have upon solid-waste generation rates.

5. To determine if any regional variations are noticed
in solid-waste generation rates.

SPECIFIC CONCLUSIONS

What conclusions were reached? The analysis of the
data from the 1968 National Survey of Community Solid Waste
Practices indicates that the average rates of solid-waste generations are significantly higher than were reported by the
U. S. Environmental Protection Agency. This includes generation rates by waste source and community. These conclusions
certainly were substantiated by the per capita generation rates
derived from the sample survey and the test case with the
Independent Virginia Cities.

In the development of the generation rates from the
sample, no regional variation in solid-waste generation rate
was noticed. This is possibly due to the small number of
sample cases for any one area or region.

Significant variations exist in generation rates
relative to the influence of socio-economic factors. The
generation rates may vary as follows:

a) Solid-waste generation increases directly with city
size.

b) Solid-waste generation increases directly with city
density.

c) Solid-waste generation increases <u>inversely</u> with income levels.

The net effect of these direct and inverse relationships varies considerably however. Thus, while solid-waste generation increases inversely with income levels, the net effect upon the total amount of solid waste generated may not be an actual reduction in waste but rather show a smaller percentage increase.

These results however are limited by the Survey data itself. The greatest limitation of the data was the unquestionably high number of case records derived from cities under 20,000 population.

APPENDIX A

NATIONAL SURVEY OF COMMUNITY SOLID
WASTE PRACTICES: QUESTIONNAIRES

DEPARTMENT OF
HEALTH, EDUCATION, AND WELFARE
PUBLIC HEALTH SERVICE

COMMUNITY SOLID WASTE PRACTICES

COMMUNITY DESCRIPTION REPORT

Form Approved
Budget Bureau No. 68-S-67022

| 1 |

11	12
0	1

1. STATE		2. COUNTY				3. COMMUNITY				
	2 3		4 5 6				7 8 9 10			

4. DOES THE COMMUNITY OPERATE OR EXERCISE JURISDICTION OVER ALL OR ANY PART OF A SOLID WASTE COLLECTION OR DISPOSAL SYSTEM? (Check appropriate categories)

	COLLECTION	DISPOSAL		13 14
OPERATE	☐ YES ☐ NO	☐ YES ☐ NO		15 16
EXERCISE JURISDICTION	☐ YES ☐ NO	☐ YES ☐ NO		17 18

5. IF THE COMMUNITY IS A COUNTY, DISTRICT, OR OTHER POLITICAL JURISDICTION CONTAINING INCORPORATED PLACES, LIST SUCH INCORPORATED PLACES INCLUDED IN THIS COMMUNITY

a. _____ 19 20 21 22 b. _____ 23 24 25 26 c. _____ 27 28 29 30

IF ADDITIONAL ENTRIES ARE REQUIRED, CHECK HERE ☐ (31) AND MAKE ADDITIONAL ENTRIES IN ITEM #39.

6. IF THE COMMUNITY IS A COUNTY, DISTRICT, OR OTHER POLITICAL JURISDICTION CONTAINING INCORPORATED PLACES, LIST SUCH INCORPORATED PLACES THAT HAVE BEEN OR WILL BE SURVEYED SEPARATELY

a. _____ 32 33 34 35 b. _____ 36 37 38 39 c. _____ 40 41 42 43

IF ADDITIONAL ENTRIES ARE REQUIRED, CHECK HERE ☐ (44) AND MAKE ADDITIONAL ENTRIES IN ITEM #39.

7. POPULATION OF COMMUNITY, 1960 CENSUS.							8. EST. CURRENT POPULATION							9. TOTAL AREA OF COMMUNITY (Sq. miles)			
45	46	47	48	49	50	51		52	53	54	55	56	57	58	59	60	61	62

10. WHICH PLANNING AGENCIES INCLUDE SOLID WASTES AS PART OF THEIR COMPREHENSIVE PLANNING? (Check all appropriate categories)

63 ☐ NO AGENCY 64 ☐ LOCAL 65 ☐ COUNTY 66 ☐ REGIONAL

S T O R A G E

11. LEGISLATIVE-ADMINISTRATIVE REGULATIONS GOVERNING ON-SITE STORAGE (Check appropriate categories)

SOURCE	GARBAGE				OTHER REFUSE				Do not use
	REGULATIONS		ENFORCED		REGULATIONS		ENFORCED		
	YES	NO	YES	NO	YES	NO	YES	NO	
HOUSEHOLD	☐	☐	☐	☐	☐	☐	☐	☐	67 68
COMMERCIAL	☐	☐	☐	☐	☐	☐	☐	☐	69 70
INDUSTRIAL	☐	☐	☐	☐	☐	☐	☐	☐	71 72
AGRICULTURAL	☐	☐	☐	☐	☐	☐	☐	☐	73 74
INSTITUTIONAL	☐	☐	☐	☐	☐	☐	☐	☐	75 76

12. LEGISLATIVE-ADMINISTRATIVE REGULATIONS FOR ON-SITE STORAGE ENFORCED PRINCIPALLY BY (Check one only)

☐ NO ENFORCEMENT
☐ OPERATIONAL AUTHORITY
☐ POLICE
☐ HEALTH AUTHORITY
☐ OTHER _____ (Specify)

| | 77 |

C O L L E C T I O N

13. SUPERVISION FOR WORK PERFORMANCE OF PRIVATE COLLECTORS PROVIDED PRIMARILY BY (Check one only)

☐ NONE ☐ HEALTH DEPARTMENT 78
☐ PUBLIC WORKS DEPARTMENT ☐ OTHER _____ (Specify)

14. COLLECTION WORK PERFORMED BY (Estimate to nearest 10% volume for each type of refuse. Rows should add horizontally to 100%)

0	2
13	14

SOURCE	PUBLIC AGENCY		PRIVATE COLLECTOR		INDIVIDUAL	
HOUSEHOLD						
	15	16	17	18	19	20
COMMERCIAL						
	21	22	23	24	25	26
INDUSTRIAL						
	27	28	29	30	31	32
INSTITUTIONAL						
	33	34	35	36	37	38
DEAD ANIMALS						
	39	40	41	42	43	44
ABANDONED VEHICLES						
	45	46	47	48	49	50

15. CLASSES OF HOUSEHOLD REFUSE NOT COLLECTED (Check appropriate categories)

GARBAGE ☐
RUBBISH ☐
YARD REFUSE ☐
ASHES ☐
COMBUSTIBLES ☐
NON-COMBUSTIBLES ☐
BULKY ITEMS (Refrigerators, etc.) ☐
OTHER (Specify) ☐

16. HOUSEHOLD REFUSE COLLECTION FREQ.

TYPES OF REFUSE COLLECTED SEPARATELY	1 PER WEEK	2 PER WEEK	OTHER	
COMBINED COLLECTION	☐	☐	☐	51
GARBAGE	☐	☐	☐	52
RUBBISH	☐	☐	☐	53
YARD REFUSE	☐	☐	☐	54
ASHES	☐	☐	☐	55
COMBUSTIBLES	☐	☐	☐	56
NON-COMBUSTIBLES	☐	☐	☐	57
BULKY ITEMS (Refrigerators, etc.)	☐	☐	☐	58
OTHER (Specify)	☐	☐	☐	59 60
OTHER (Specify)	☐	☐	☐	61 62
OTHER (Specify)	☐	☐	☐	63 64

COMMUNITY DESCRIPTION REPORT (Page 2)

19. AVERAGE MANPOWER (MAN-YEARS) AND EQUIPMENT USED FOR COLLECTING COMMUNITY SOLID WASTES	0 3
	13 14

COLLECTION

17. DEAD ANIMALS COLLECTED ANNUALLY		FIGURES AT LEFT REFER TO				SOURCES OF WASTES										
						HOUSEHOLD COMMERCIAL INSTITUTIONAL				STREET CLEANING				INDUSTRIAL		
		NUMBER OF ANIMALS	TONS OF ANIMALS	ARE THESE FIGURES ESTIMATES?	P U B L I C	COLLECTORS AND DRIVERS (Man-years)	15	16 17	18	19 20	21 22		23	24 25		
						COMPACTOR TRUCKS (Number)	26	27 28	29		30 31			32 33		
LARGE ANIMALS	☐	☐	☐ YES		OTHER VEHICLES (Number)	34	35 36		37	38 39			40 41		
65 66 67		68	☐ NO												
SMALL ANIMALS	☐	☐	☐ YES	P R I V A T E	NUMBER OF FIRMS	42	43 44	45		46 47	48 49	50 51			
			☐ NO		COLLECTORS AND DRIVERS (Man-years)	52	53 54	55		56 57	58 59	60 61			
69 70 71 72 73 74		75			COMPACTOR TRUCKS (Number)	62	63 64	65		66 67	68 69	70			
18. NUMBER OF ABANDONED VEHICLES COLLECTED ANNUALLY					OTHER VEHICLES (Number)	71	72 73	74		75 76	77 78	79 80			
(Enter number) 76 77 78 79 80															

20. INDICATE AMOUNTS (MEASURED OR ESTIMATED) OF COMMUNITY SOLID WASTES COLLECTED ANNUALLY

CLASSIFICATION		MEASURED	ESTIMATED	TONS	YARDS	
REFUSE (Household)	0 4			☐	☐	
	13 14	15 16 17 18 19 20 21	22 23 24 25 26 27 28			29
REFUSE (Commercial)				☐	☐	
		30 31 32 33 34 35 36	37 38 39 40 41 42 43			44
REFUSE (Combined household and commercial)				☐	☐	
		45 46 47 48 49 50 51	52 53 54 55 56 57 58			59
REFUSE (Industrial)	0 5			☐	☐	
	13 14	60 61 62 63 64 65 66	67 68 69 70 71 72 73			74
REFUSE (Agricultural)				☐	☐	
		15 16 17 18 19 20 21	22 23 24 25 26 27 28			29
REFUSE (Institutional)				☐	☐	
		30 31 32 33 34 35	36 37 38 39 40 41			42
DEMOLITION AND CONSTRUCTION REFUSE				☐	☐	
		43 44 45 46 47 48	49 50 51 52 53 54			55
STREET AND ALLEY CLEANINGS				☐	☐	
		56 57 58 59 60	61 62 63 64 65			66
TREE AND LANDSCAPING REFUSE				☐	☐	
		67 68 69 70 71	72 73 74 75 76			77
PARK AND BEACH REFUSE	0 6			☐	☐	
	13 14	15 16 17 18 19	20 21 22 23 24			25
CATCH BASIN REFUSE				☐	☐	
		26 27 28 29 30	31 32 33 34 35			36
SEWAGE TREATMENT PLANT SOLIDS AND PUMPING STATION CLEANINGS				☐	☐	
		37 38 39 40 41	42 43 44 45 46			47

DISPOSAL

21. AGENCY PRIMARILY RESPONSIBLE FOR RECOMMENDATIONS FOR LOCATION AND DEVELOPMENT OF NEW DISPOSAL SITES OR FACILITIES (Check one only)
☐ NONE ☐ HEALTH AUTHORITY
☐ OPERATIONAL AUTHORITY ☐ OTHER (Specify) _____
☐ 48

22. AGENCY PRIMARILY RESPONSIBLE FOR REGULATION OF DISPOSAL FACILITY OPERATIONS (Check one only)
☐ NONE ☐ POLICE ☐ OTHER (Specify)
☐ OPERATIONAL AUTHORITY ☐ HEALTH AUTHORITY _____
☐ 49

	PRACTICED?	REGULATED?	
23. IS ON-SITE OPEN BURNING OF DEMOLITION, CONSTRUCTION, AND/OR LAND CLEARANCE WASTES . . .	☐ YES ☐ NO	☐ YES ☐ NO	50 51
24. IS BACKYARD BURNING OF HOUSEHOLD REFUSE . . .	☐ YES ☐ NO	☐ YES ☐ NO	52 53
25. IS ON-SITE OPEN BURNING OF COMMERCIAL, INSTITUTIONAL, INDUSTRIAL AND/OR AGRICULTURAL WASTES . . .	☐ YES ☐ NO	☐ YES ☐ NO	54 55

COMMUNITY DESCRIPTION REPORT (PAGE 3)

DISPOSAL

24. INDICATE BELOW THE NUMBER OF REDUCTION AND/OR DISPOSAL SITES UTILIZED BY THE COMMUNITY'S PUBLIC COLLECTORS, PRIVATE COLLECTORS AND/OR INDIVIDUAL HAULERS *(Enter Numbers)*

SITES	PUBLICLY OPERATED		PRIVATELY OPERATED	
LAND DISPOSAL SITES		56	57	58	59
INCINERATORS	60	61	62	63
TRANSFER STATIONS	64	65	66	67
HOG FEEDING LOTS	68	69	70	71
COMPOST PLANTS		72		73
TEPEE BURNERS	74	75	76	77
OTHER *(Specify)*		78		79

Do not use 60 13 14 | 0 | 7 |

27. REDUCTION AND/OR DISPOSAL SITES SERVING THE COMMUNITY

TOTAL			NUMBER OUTSIDE COMMUNITY	
	15	16	19	20
NUMBER WITHIN COMMUNITY	17	18		

26. USE OF COMPLETED LAND DISPOSAL SITES

(List no. of sites in each category as applicable)

COMPLETED SITE USE	NUMBER	
RECREATIONAL AREA OR PARK	21	22
PARKING LOT	23	24
LIGHT CONSTRUCTION	25	26
HEAVY CONSTRUCTION	27	28
AGRICULTURE	29	30
NO SPECIFIC USE	31	32
OTHER *(Specify)*	33	34
	Do not use 35	
OTHER *(Specify)*	36	37
	Do not use 38	
OTHER *(Specify)*	39	40
	Do not use 41	

29 NUMBER OF PROMISCUOUS DUMPS WITHIN THE COMMUNITY'S BOUNDARIES KNOWN TO BE ACTIVE

(Enter Number) | 42 | 43 |

IF INFORMATION IS NOT AVAILABLE CHECK HERE ☐

30 ESTIMATED NUMBER OF HOUSEHOLD GARBAGE GRINDERS INSTALLED

(Enter Number) | 44 | 45 | 46 | 47 | 48 | 49 |

31. ESTIMATED NUMBER OF GARBAGE GRINDERS IN COMMERCIAL AND INSTITUTIONAL ESTABLISHMENTS

(Enter Number) | 50 | 51 | 52 | 53 | 54 |

32. ESTIMATED NUMBER OF ON-SITE INCINERATORS SERVING APARTMENT HOUSES, COMMERCIAL AND INSTITUTIONAL ESTABLISHMENTS

(Enter Number) | 55 | 56 | 57 | 58 | 59 |

33. ESTIMATED NUMBER OF HOUSEHOLD ON-SITE INCINERATORS

(Enter Number) | 60 | 61 | 62 | 63 | 64 |

BUDGET (FISCAL)

34. COMMUNITY FUNDS BUDGETED FOR COLLECTION OF SOLID WASTES FOR CALENDAR OR FISCAL YEAR, 1967

a. EXCLUDING CAPITAL EXPENDITURES

$ | 65 | 66 | 67 | 68 | 69 | 70 | 71 | 72 |

b. CAPITAL EXPENDITURES ONLY

$ | 73 | 74 | 75 | 76 | 77 | 78 | 79 | 80 |

35. COMMUNITY FUNDS BUDGETED FOR DISPOSAL OF SOLID WASTES FOR CALENDAR OR FISCAL YEAR, 1967 | 0 | 8 | 13 | 14 |

a. EXCLUDING CAPITAL EXPENDITURES

$ | 15 | 16 | 17 | 18 | 19 | 20 | 21 | 22 |

b. CAPITAL EXPENDITURES ONLY

$ | 23 | 24 | 25 | 26 | 27 | 28 | 29 | 30 |

36. CLASSES OF REFUSE FOR WHICH COLLECTION AND/OR DISPOSAL FUNDS HAVE BEEN BUDGETED FOR CALENDAR OR FISCAL YEAR, 1967

REFUSE CATEGORY	COLLECTION	DISPOSAL
REFUSE *(Household)*	31 ☐	32 ☐
REFUSE *(Commercial)*	33 ☐	34 ☐
REFUSE *(Industrial)*	35 ☐	36 ☐
REFUSE *(Agricultural)*	37 ☐	38 ☐
REFUSE *(Institutional)*	39 ☐	40 ☐
STREET AND ALLEY CLEANINGS	41 ☐	42 ☐
DEMOLITION AND CONSTRUCTION REFUSE	43 ☐	44 ☐
TREE AND LANDSCAPING REFUSE	45 ☐	46 ☐
PARK AND BEACH REFUSE	47 ☐	48 ☐
CATCH BASIN REFUSE	49 ☐	50 ☐
SEWAGE TREATMENT/PLANT SOLIDS AND PUMPING STATION CLEANINGS	51 ☐	52 ☐
DEAD ANIMALS	53 ☐	54 ☐
ABANDONED VEHICLES	55 ☐	56 ☐
OTHER *(Specify)*	57 ☐	58 ☐
OTHER *(Specify)*	60 ☐	61 ☐
OTHER *(Specify)*	63 ☐	64 ☐

59 ☐
42 ☐
45 ☐

COMMUNITY DESCRIPTION REPORT (Page 4)

37. NAME AND TITLE OF PERSON COMPLETING FORM	ORGANIZATION AND ADDRESS	DATE OF SURVEY		
		DAY	MONTH	YEAR
		66 67	68 69	70 71

38. IF SOURCES OTHER THAN REPORTER DESIGNATED IN ITEM 37 WERE UTILIZED IN COMPLETING THIS FORM, INDICATE BELOW THE SOURCES USED AND ITEM NUMBER(S)

D A T A S O U R C E S	NAME OF PERSON	TITLE	ORGANIZATION	ITEM NUMBER(S)

39. CONTINUATION ITEMS

ITEM NO.	ADDITIONAL INFORMATION

40. REMARKS (Attach additional sheet if necessary)

DEPARTMENT OF
HEALTH, EDUCATION, AND WELFARE
PUBLIC HEALTH SERVICE

Form Approved
Budget Bureau No. 68-S-67022

COMMUNITY SOLID WASTE PRACTICES
LAND DISPOSAL SITE INVESTIGATION REPORT

| 2 |

1. STATE | 2 3 | **2. COUNTY** | 4 5 6 | **3. SITE LOCATION** *(Political Jurisdiction)* | 7 8 9 10

4. NAME OF SITE | 11 12 13 | **5. ADDRESS OF SITE** | 1 | **6. DATE OF SURVEY** DAY MONTH YEAR | 14 | 15 16 17 18 19 20

7. NAME OF PERSON COMPLETING FORM | **8. TITLE** | **9. ORGANIZATION AND ADDRESS**

10. POLITICAL JURISDICTIONS SERVED BY LAND DISPOSAL SITE

NAME OF POLITICAL JURISDICTION	ESTIMATED PERCENTAGE OF JURISDICTION SERVED BY SITE	AVERAGE DISTANCE OF SITE FROM CENTER OF SOURCE AREA *(Miles)*
	21 22 23 24	25 26	27 28
	29 30 31 32	33 34	35 36
	37 38 39 40	41 42	43 44
	45 46 47 48	49 50	51 52

FOR ADDITIONAL ENTRIES, CHECK HERE ☐ (53) AND MAKE ENTRIES IN ITEM #45

11. SITE OPERATED BY
☐ PUBLIC AGENCY
☐ PRIVATE AGENCY | 54

12. SITE OWNED BY
☐ PUBLIC AGENCY
☐ PRIVATE AGENCY | 55

13. IS OPERATION REGULATED BY A HEALTH AUTHORITY? ☐ YES ☐ NO | IF YES, INDICATE LEVEL OF PRINCIPAL AUTHORITY *(Check one only)* ☐ COMMUNITY ☐ COUNTY ☐ STATE ☐ OTHER _____ *(Specify)* | 56 57

14. GENERAL CHARACTER OF SITE *(Check one only)*
☐ QUARRY OR BORROW PIT
☐ GULLY-CANYON
☐ LEVEL AREAS
☐ OTHER _____ *(Specify)*
☐ HILLSIDE
☐ MARSH, TIDELAND OR FLOOD PLAIN

Do not use 58

15. YEAR SITE PLACED IN OPERATION 19 | 59 60
16. ANTICIPATED LIFE REMAINING *(Years)* | 61 62 63
17. TOTAL AREA OF SITE *(Acres)* | 64 65 66 67
18. AREA TO BE USED FOR LAND DISPOSAL *(Acres)* | 68 69 70 71

19. ZONING/LAND USE SURROUNDING FACILITY *(Check predominant type only)*

ZONING
☐ NONE
☐ RESIDENTIAL
☐ COMMERCIAL
☐ INDUSTRIAL
☐ AGRICULTURAL
☐ OTHER _____ *(Specify)*

LAND USE
☐ RESIDENTIAL
☐ COMMERCIAL
☐ INDUSTRIAL
☐ AGRICULTURAL
☐ OTHER _____ *(Specify)* | 72 73

20. IS USE OF COMPLETED SITE PLANNED? ☐ YES ☐ NO | IF YES, CHECK PREDOMINANT USE ONLY ☐ RECREATIONAL AREA OR PARK ☐ PARKING LOT ☐ LIGHT CONSTRUCTION ☐ HEAVY CONSTRUCTION ☐ AGRICULTURE ☐ OTHER _____ *(Specify)* ☐ USE NOT DETERMINED | 74

21. WILL PUBLIC AGENCY CONTROL COMPLETED SITE USE? ☐ YES ☐ NO | **22. MATERIAL USED FOR COVER** *(Check one only)* ☐ NONE ☐ EARTH ☐ OTHER _____ *(Specify)* | 75 76

23. FREQUENCY OF COVER *(Check one only)* ☐ NONE ☐ DAILY *(End of each working day)* ☐ DAILY *(Except face)* ☐ OTHER _____ *(Specify)* | **24. IS SPREADING AND COMPACTION OF REFUSE HANDLED IN APPROX-IMATELY TWO-FOOT LAYERS OR LESS?** ☐ YES ☐ NO | 77 78

25. NUMBER OF DAYS DISPOSAL SITE COULD NOT BE USED BECAUSE OF WEATHER CONNECTED CONDITIONS *(Enter average per year)* | 79 80

26. GENERAL CHARACTER OF OPERATION *(Judgment evaluation - check appropriate categories)* | 2 | 14

APPEARANCE	IS BLOWING PAPER CONTROLLED?	IS BLOWING PAPER CONSIDERED TO BE A NUISANCE?	ROUTINE BURNING	ARE THERE SURFACE DRAINAGE PROBLEMS?	ARE THERE LEACHING PROBLEMS?
☐ SIGHTLY ☐ UNSIGHTLY	☐ YES ☐ NO	☐ YES ☐ NO	☐ NONE ☐ UNCONTROLLED ☐ PLANNED AND LIMITED	☐ YES ☐ NO	☐ YES ☐ NO
15	16	17	18	19	20

LAND DISPOSAL SITE INVESTIGATION REPORT (Page 2)

27. CONTROL PROGRAMS

		YES	NO	Do not use
RODENT CONTROL PROGRAM	NEEDED	☐	☐	21
	PROVIDED	☐	☐	22
FLY CONTROL PROGRAM	NEEDED	☐	☐	23
	PROVIDED	☐	☐	24
BIRD CONTROL PROGRAM	NEEDED	☐	☐	25
	PROVIDED	☐	☐	26
DUST CONTROL PROGRAM	NEEDED	☐	☐	27
	PROVIDED	☐	☐	28
ODOR CONTROL PROGRAM	NEEDED	☐	☐	29
	PROVIDED	☐	☐	30

28. IS LOWEST PART OF FILL IN WATER TABLE? ☐ YES ☐ NO   ☐ 31

29. FIRE PROTECTION ☐ NONE ☐ WATER ☐ FIREBREAK ☐ OTHER ___ (Specify)   ☐ 32

30. NUMBER OF TIMES FIRE CONTROL EQUIPMENT WAS REQUIRED AT SITE IN THE PAST YEAR [ ][ ][ ] 33 34 35

31. IS SALVAGING PERMITTED? ☐ YES ☐ NO   ☐ 36

32. IS SALVAGING PRACTICED? ☐ YES ☐ NO   ☐ 37

33. ESTIMATED NUMBER OF LOADS DEPOSITED DAILY (Average)

FROM PUBLIC COLLECTION VEHICLES (Enter number) [ ] 38 39 40	FROM PRIVATE COLLECTION VEHICLES (Enter number) [ ] 41 42 43	FROM OTHER VEHICLES (Specify) (Enter number) [ ] 44 45 46

34. ARE QUANTITATIVE RECORDS KEPT IN ANY FORM? ☐ YES ☐ NO Do not use 47

35. QUANTITIES OF SOLID WASTES RECEIVED ANNUALLY

TONS WEIGHED		48 49 50 51 52 53 54
TONS ESTIMATED		55 56 57 58 59 60 61
CUBIC YARDS		62 63 64 65 66 67 68 69

36. GENERAL CLASSIFICATION OF SOLID WASTES ACCEPTED AT DISPOSAL SITE (Check those accepted)

☐ HOUSEHOLD 70 ☐ INDUSTRIAL 72 ☐ INSTITUTIONAL 74
☐ COMMERCIAL 71 ☐ AGRICULTURAL 73 ☐ INCINERATOR RESIDUE ONLY 75

37. CHECK ANY ITEMS LISTED BELOW WHICH ARE EXCLUDED FROM THE DISPOSAL SITE   3 14

☐ ALL PUTRESCIBLES 15 ☐ SEWAGE SOLIDS 21 ☐ TIRES 27
☐ ALL NON-COMBUSTIBLES 16 ☐ JUNKED AUTOMOBILES 22 ☐ HAZARDOUS MATERIALS 28
☐ ALL COMBUSTIBLES 17 ☐ LARGE APPLIANCES 23 ☐ OTHER (Specify) 29
☐ GARBAGE 18 ☐ DEMOLITION WASTES 24 ☐ OTHER (Specify) 31   ☐ 30
☐ DEAD ANIMALS 19 ☐ CONSTRUCTION DEBRIS 25 ☐ OTHER (Specify) 32
☐ WASTE OIL 20 ☐ STREET SWEEPINGS 26 ☐ OTHER (Specify) 33   ☐ 34

38. EQUIPMENT AVAILABLE (Average utilized daily)

	NUMBER
DRAGLINE OR SHOVEL-TYPE EXCAVATORS	35 35
SCRAPERS (Self-propelled)	37 38
TRACTORS (Track or Rubber Tire) (Bulldozer or High Lift Loader)	39 40
TRUCKS	41 42
OTHER ___ (Specify)	Do not use 43 44 45
OTHER ___ (Specify)	Do not use 46 47 48

39. TOTAL NUMBER OF EMPLOYEES ON SITE (Average daily) [ ] 49 50

40. HOURS OF DAILY OPERATION (On a 24-hour clock) BEGIN [ ] 51 52 END [ ] 53 54

41. NUMBER OF DAYS OPERATED PER WEEK [ ] 55

42. ANNUAL OPERATING COST (Including supervision and equipment maintenance) $ [ ] 56 57 58 59 60 61 62

43. IS THIS A SANITARY LANDFILL? ☐ YES ☐ NO   ☐ 63

44. IF SOURCES OTHER THAN REPORTER DESIGNATED IN ITEM 7 WERE UTILIZED IN COMPLETING THIS FORM, INDICATE BELOW THE SOURCES USED AND ITEM NUMBERS

NAME OF PERSON	TITLE	ORGANIZATION	ITEM NUMBER(S)

LAND DISPOSAL SITE INVESTIGATION REPORT (Page 3)

45. CONTINUATION ITEMS

ITEM NO.	ADDITIONAL INFORMATION

46. REMARKS (Attach additional sheet if necessary)

DEPARTMENT OF
HEALTH, EDUCATION, AND WELFARE
PUBLIC HEALTH SERVICE

COMMUNITY SOLID WASTE PRACTICES

FACILITY INVESTIGATION REPORT

Form Approved
Budget Bureau No. 68-S-67022

3

1. STATE

2. COUNTY

3. FACILITY LOCATION (Political jurisdiction)

4. NAME OF FACILITY

5. ADDRESS OF FACILITY

6. DATE OF SURVEY
DAY   MONTH   YEAR

7. TYPE OF FACILITY (Check one)
☐ INCINERATOR   ☐ COMPOST   ☐ TEPEE BURNER
☐ TRANSFER STATION   ☐ HOG FEEDING   ☐ OTHER _____ (Specify)

21

8. NAME OF PERSON COMPLETING FORM

9. TITLE

10. ORGANIZATION AND ADDRESS

11. POLITICAL JURISDICTION SERVED BY FACILITY

NAME OF POLITICAL JURISDICTION	ESTIMATED PERCENTAGE OF JURISDICTION SERVED BY FACILITY	AVERAGE DISTANCE OF FACILITY FROM CENTER OF SOURCE AREA (Miles)
	22 23 24 25	26 27	28 29
	30 31 32 33	34 35	36 37
	38 39 40 41	42 43	44 45
	46 47 48 49	50 51	52 53

FOR ADDITIONAL ENTRIES, CHECK HERE ☐ AND MAKE ENTRIES IN ITEM # 48

12. FACILITY OPERATION AND OWNERSHIP
a. OPERATED BY
☐ PUBLIC AGENCY
☐ PRIVATE AGENCY
b. OWNED BY
☐ PUBLIC AGENCY
☐ PRIVATE AGENCY

55

13. YEAR FACILITY WAS CONSTRUCTED TO PRESENT CAPACITY
19 ___ 56 57

14. IS OPERATION REGULATED BY A HEALTH AUTHORITY?
☐ YES   ☐ NO
IF YES, INDICATE LEVEL OF PRINCIPAL AUTHORITY (Check one only)
☐ COMMUNITY   ☐ STATE
☐ COUNTY   ☐ OTHER _____ (Specify)

58

15. ZONING LAND USE SURROUNDING FACILITY (Check predominant type only)
ZONING
☐ NONE   ☐ INDUSTRIAL
☐ RESIDENTIAL   ☐ AGRICULTURAL
☐ COMMERCIAL   ☐ OTHER _____ (Specify)
LAND USE
☐ RESIDENTIAL   ☐ AGRICULTURAL
☐ COMMERCIAL   ☐ OTHER _____ (Specify)
☐ INDUSTRIAL

59 60

16. CAPACITY OF FACILITY (Tons per 24 hours)
AVERAGE DAILY INPUT 61 62 63
RATED CAPACITY 64 65 66 67

17. NUMBER OF HOURS FACILITY OPERATED (Average daily) 68 69

18. NUMBER OF DAYS FACILITY OPERATED PER WEEK 70

19. IF FACILITY IS A HOG FEEDING ESTABLISHMENT, IS GARBAGE COOKED? ☐ YES ☐ NO

71

20. ESTIMATED NUMBER OF LOADS DEPOSITED DAILY (Average)
FROM PUBLIC COLLECTION VEHICLES 72 73 74
FROM PRIVATE COLLECTION VEHICLES 75 76 77
FROM OTHER VEHICLES 78 79 80

21. GENERAL CHARACTER OF OPERATION (Judgment evaluation—check appropriate categories)

2
14

APPEARANCE	IS BLOWING PAPER CONTROLLED?	IS BLOWING PAPER CONSIDERED TO BE A NUISANCE?	ARE THERE SURFACE DRAINAGE PROBLEMS?	ARE THERE LEACHING PROBLEMS?
☐ SIGHTLY ☐ UNSIGHTLY 15	☐ YES ☐ NO 16	☐ YES ☐ NO 17	☐ YES ☐ NO 18	☐ YES ☐ NO 19

22. CONTROL PROGRAMS (Judgment evaluation—check appropriate categories)

PROGRAM		YES	NO	Do not use	PROGRAM		YES	NO	Do not use
RODENT CONTROL PROGRAM	NEEDED	☐	☐	20	DUST CONTROL PROGRAM	NEEDED	☐	☐	26
	PROVIDED	☐	☐	21		PROVIDED	☐	☐	27
FLY CONTROL PROGRAM	NEEDED	☐	☐	22	ODOR CONTROL PROGRAM	NEEDED	☐	☐	28
	PROVIDED	☐	☐	23		PROVIDED	☐	☐	29
BIRD CONTROL PROGRAM	NEEDED	☐	☐	24	WASTE WATER EFFLUENT TREATMENT	NEEDED	☐	☐	30
	PROVIDED	☐	☐	25		PROVIDED	☐	☐	31

FACILITY INVESTIGATION REPORT (Page 2)

23. ARE QUANTITATIVE RECORDS KEPT IN ANY FORM ☐ YES ☐ NO `32`

24. QUANTITIES OF SOLID WASTES RECEIVED ANNUALLY	25. GENERAL CLASSIFICATION OF SOLID WASTES ACCEPTED AT DISPOSAL FACILITY (Check those accepted)

TONS WEIGHED ☐☐☐☐☐☐☐ `33 34 35 36 37 38 39`

TONS ESTIMATED ☐☐☐☐☐☐☐ `40 41 42 43 44 45 46`

CUBIC YARDS ☐☐☐☐☐☐☐☐ `47 48 49 50 51 52 53 54`

☐ HOUSEHOLD `55`   ☐ AGRICULTURAL `58`
☐ COMMERCIAL `56`   ☐ INSTITUTIONAL `59`
☐ INDUSTRIAL `57`

26. CHECK ANY ITEMS LISTED BELOW WHICH ARE EXCLUDED FROM THE DISPOSAL FACILITY

☐ ALL PUTRESCIBLES `60`   ☐ SEWAGE SOLIDS `64`   ☐ CONSTRUCTION DEBRIS `68`   ☐ HAZARDOUS MATERIAL `72`

☐ ALL NON-COMBUSTIBLES `61`   ☐ JUNKED AUTOMOBILES `65`   ☐ STREET SWEEPINGS `69`   ☐ OTHER _____ `73` (Specify) `74`

☐ GARBAGE `62`   ☐ LARGE APPLIANCES `66`   ☐ TIRES `70`   ☐ OTHER _____ `75` (Specify) `76`

☐ DEAD ANIMALS `63`   ☐ DEMOLITION WASTES `67`   ☐ WASTE OIL `71`   ☐ OTHER _____ `77` (Specify) `78`

27. IS RESIDUE DISPOSAL IN A LAND DISPOSAL SITE? ☐ NO RESIDUE ☐ YES ☐ NO ☐ OTHER DISPOSAL _____ (Specify) `3`

28. IF RESIDUE DISPOSAL IS IN A LANDFILL, COMPLETE THE FOLLOWING: `14 15`

DISTANCE TO SITE (Miles) `16 17`	IS RESIDUE COMPACTED? ☐ YES ☐ NO `18`	IS RESIDUE COVERED WITH EARTH? ☐ YES ☐ NO `19`

29. IS SALVAGING PERMITTED? ☐ YES ☐ NO `20`   **30. IS SALVAGING PRACTICED?** ☐ YES ☐ NO `21`   **31. TOTAL NUMBER OF EMPLOYEES ON SITE (Average daily)** `22 23 24`

32. ESTIMATED REPLACEMENT COST OF FACILITY (Excluding land cost) $ ☐☐☐☐☐☐☐☐ `25 26 27 28 29 30 31 32`   **33. ANNUAL OPERATING COST (Including supervision and maintenance)** $ ☐☐☐☐☐☐☐ `33 34 35 36 37 38 39`

34. ANNUAL REVENUE (IF ANY) FROM SALVAGE $ ☐☐☐☐☐ `40 41 42 43 44`   **35. TOTAL ANNUAL REVENUE** $ ☐☐☐☐☐☐ `45 46 47 48 49 50`

36. TOTAL NUMBER OF FURNACES `51`   **37. NUMBER OF FURNACES OPERATED (Average daily)** `52`   **38. PIT VOLUME (Cubic yards)** `53 54 55 56 57`

39. NUMBER OF STACKS `58`   **40. METHOD OF CHARGING** ☐ BATCH ☐ CONTINUOUS ☐ OTHER _____ (Specify) `59`

41. RESIDUE RECLAMATION

METALS SALVAGED ☐ YES ☐ NO `60`   SPECIAL PURPOSE USE OF RESIDUE (Check appropriate category) ☐ COVER MATERIAL ☐ ROAD CONSTRUCTION ☐ OTHER _____ (Specify) `61`

42. PERCENTAGE REDUCTION OF REFUSE

BY WEIGHT `62 63`   BY VOLUME `64 65`   NOTE: PERCENTAGE REDUCTION $= \dfrac{INPUT-RESIDUE}{INPUT} \times 100$

(Left margin: **I N C I N E R A T O R S   O N L Y**)

43. STACK EMISSION CONTROL EQUIPMENT	UTILIZATION		PERFORMANCE		Do Not Use
	AVAILABLE	IN USE	SATISFACTORY	UNSATISFACTORY	
DRY (NON-MECHANICAL)	☐	☐	☐	☐	`66 67`
WATER (SPRAYS, SCRUBBERS, ETC.)	☐	☐	☐	☐	`68 69`
MECHANICAL (CYCLONES, FILTERS, ETC.)	☐	☐	☐	☐	`70 71`
ELECTRICAL (PRECIPITATORS, ETC.)	☐	☐	☐	☐	`72 73`
OTHER _____ (Specify)	☐	☐	☐	☐	`74 75`

`76`

44. IS STACK PLUME OBSERVABLE? ☐ YES ☐ NO `77`   **45. EQUIPMENT (Average daily number of vehicles used for residue hauling)** `78 79`   **46. IS WASTE HEAT RECLAMATION PRACTICED?** ☐ YES ☐ NO `80`

FACILITY INVESTIGATION REPORT (Page 3)

47. IF SOURCES OTHER THAN REPORTER DESIGNATED IN ITEM 8 WERE UTILIZED IN COMPLETING THIS FORM, INDICATE BELOW THE SOURCES USED AND ITEM NUMBERS

NAME OF PERSON.	TITLE	ORGANIZATION	ITEM NUMBER(S)

48. CONTINUATION ITEMS

ITEM NO.	ADDITIONAL INFORMATION

49. REMARKS *(Attach additional sheet if necessary)*

APPENDIX B

LIST OF COUNTIES AND STATES FOR WHICH
SOCIO-ECONOMIC DATA WAS COLLECTED

APPENDIX B

tate Code	State	# of Cases per county (Blank = 1)	County Code	County
01	Alabama		001	Autauga
			005	Clarke
			051	Elmore
			071	Jackson
			079	Laurence
			099	Monroe
			123	Tallapoosa
03	Arkansas		009	Boone
			037	Cross
			069	Jefferson
			105	Perry
			131	Sebastian
04	California		001	Almeda
			013	Contra Costa
			019	Fresno
			029	Kern
		(2)	037	Los Angeles
			039	Madera
			051	Mono
			059	Orange
			065	Riverside
			071	San Bernadino
			081	San Mateo
			085	Santa Clara
			103	Tehama
06	Connecticut		001	Fairfield
			003	Hartford
			005	Litchfield
			009	New Haven
			011	New London
			015	Windham
07	Delaware		005	Sussex
09	Florida		009	Brevard
			025	Dade
			047	Hamilton
			073	Leon
			095	Orange
			105	Polk
			131	Walton

State Code	State	# of Cases per county (Blank = 1)	County Code	County
10	Georgia		051	Chatham
			089	De Kalb
			137	Habersham
			243	Randolph
			313	Whitfield
11	Idaho		051	Jefferson
12	Illinois		043	Du Page
13	Indiana		021	Clay
			051	Gibson
			081	Johnson
			097	Marion
14	Iowa		015	Boone
			087	Henry
			131	Mitchell
			179	Wapello
15	Kansas		017	Chase
			057	Ford
			091	Johnson
			123	Mitchell
			155	Reno
			187	Stanton
16	Kentucky		029	Bullitt
			089	Greenup
			151	Madison
17	Louisiana		001	Acadia
			015	Bossier
			031	DeSoto
			047	Iberville
			059	La Salle
			079	Rapides
			097	St. Landry
			109	Terrebonne
18	Maine		001	Androscoggin
			023	Sagadahoc
19	Maryland		015	Cecil
			033	Prince Geor

ate Code	State	# of Cases per county (Blank = 1)	County Code	County
20	Massachusetts		003	Berkshire
			005	Bristol
			009	Essex
			011	Franklin
			013	Hampden
		(2)	017	Middlesex
			021	Norfolk
			023	Plymouth
		(2)	027	Worcester
21	Michigan		017	Bay
			037	Clinton
			045	Eaton
			049	Genessee
			065	Ingham
			075	Jackson
			077	Kalamazoo
			081	Kent
			099	Macomb
			121	Muskegon
		(2)	125	Oakland
			145	Saginaw
			159	Van Buren
		(2)	163	Wayne
22	Minnesota		103	Nicollet
23	Mississippi		003	Alcorn
			013	Calhoun
			027	Coahoma
			037	Franklin
			051	Holmes
			063	Jefferson
			077	Lawrence
			089	Madison
			099	Neshoba
			113	Pike
			123	Scott
			133	Sunflower
			143	Tunica
			161	Yalobusha
24	Missouri		011	Burton
			031	Cape Girardeau
			047	Clay
			067	Douglas
			077	Greene
			095	Jackson
			105	Laclede
			117	Livingston
			135	Moniteau
			147	Nodaway
			167	Polk
			181	Ripley

State Code	State	# of Cases per county (Blank = 1)	County Code	County
24	Missouri		189	St. Louis
			197	Schuyler
			211	Sullivan
25	Montana		021	Dawson
			047	Lake
			081	Ravalli
27	Nevada		003	Clark
29	New Jersey	(3)	003	Bergen
			005	Burlington
			007	Camden
			009	Cape May
			013	Essex
			015	Gloucester
			019	Hunterdon
			021	Mercer
		(2)	025	Monmouth
		(2)	027	Morris
			029	Ocean
			033	Salem
			037	Sussex
			039	Union
			041	Warren
30	New Mexico		017	Grant
			037	Quay
			057	Torrance
31	New York		007	Broome
			015	Chemung
		(2)	029	Erie
			043	Herkimer
			045	Jefferson
			055	Monroe
			059	Nassau
			063	Niagara
			065	Oneida
			067	Onondago
			071	Orange
			075	Oswego
			083	Rensselaer
			091	Saratoga
			103	Suffolk
			111	Ulster
			119	Westchester
32	North Carolina		005	Allegheny
			031	Carteret
			057	Davidson
			079	Greene

State Code	State	# of Cases per county (Blank = 1)	County Code	County
32	North Carolina		101	Johnston
			123	Montgomery
			151	Randolph
			175	Transylvania
			191	Wayne
33	North Dakota		005	Bensen
			017	Cass
			029	Emmons
			045	LaMoure
			055	Mclean
			063	Nelson
			077	Richland
			087	Slope
			101	Ward
34	Ohio		003	Allen
			009	Athens
			013	Belmont
			017	Butler
			025	Clermont
			033	Crawford
			035	Cuyahoga
			037	Darke
			045	Fairfield
			049	Franklin
			055	Geauga
			061	Hamilton
			063	Hancock
			067	Harrison
			077	Huron
			085	Lake
			089	Licking
			093	Lorain
			101	Marion
			107	Mercer
			113	Montgomery
			119	Muskingum
			127	Perry
			133	Portage
			137	Putnam
			149	Shelly
			153	Summit
			157	Tuscarawas
			169	Wayne
			173	Wood
			175	Wyandot
35	Oklahoma		027	Cleveland
			045	Ellis
			079	Le Flore
			109	Oklahoma
			119	Payne
			141	Tillman

State Code	State	# of Cases per county (Blank = 1)	County Code	County
36	Oregon		011	Coos
			061	Union
37	Pennsylvania		001	Adams
		(4)	003	Allegheny
		(2)	005	Armstrong
			007	Beaver
		(2)	009	Bedford
		(2)	011	Berks
			013	Blair
		(2)	015	Bradford
			017	Bucks
		(2)	019	Butler
		(2)	021	Cambria
			023	Cameron
		(2)	027	Centre
		(2)	029	Chester
			031	Clarion
		(2)	033	Clearfield
			035	Clinton
			037	Columbia
		(2)	039	Crawford
			041	Cumberland
			043	Dauphin
		(2)	045	Delaware
		(2)	049	Erie
			051	Fayette
			055	Franklin
			059	Greene
		(2)	061	Huntington
			063	Indiana
			065	Jefferson
			069	Lackawanna
			071	Lancaster
			073	Lawrence
			075	Lebanon
			077	Leheigh
		(2)	079	Luzerne
		(2)	081	Lycoming
			085	Mercer
			089	Monroe
		(2)	091	Montgomery
			093	Montour
			095	Northampton
			097	Northcumberland
			099	Perry
			103	Phildelphia
			105	Potter
		(2)	107	Schuylkill
		(2)	111	Somerset
		(2)	115	Susquehanna
			117	Tioga
			119	Union
			121	Venango
		(2)	125	Washington

tate Code	State	# of Cases per county (Blank=1)	County Code	County
37	Pennsylvania		127	Wayne
		(2)	129	Westmoreland
			131	Wyoming
		(2)	133	York
38	Rhode Island		007	Providence
39	South Carolina		005	Allendale
			015	Berkeley
			025	Chesterfield
			037	Edgefield
			049	Hampton
			059	Laurens
			071	Newberry
			077	Pickins
			087	Union
40	South Dakota		001	Aurora
			031	Carson
			047	Fall River
			067	Hutchinson
			083	Lincoln
			099	Minnehaha
			115	Spink
			131	Washabaugh
41	Tennessee		015	Cannon
			031	Coffee
			043	Dickson
			053	Gibson
			069	Hardiman
			083	Houston
			101	Lewis
			115	Marion
			131	Obion
			145	Roane
			159	Smith
			175	Van Buren
42	Texas		001	Anderson
			027	Bell
			039	Brazoria
			061	Cameron
			085	Collins
			105	Crockett
			117	Deaf Smith
			139	Ellis
			159	Franklin
			179	Gray
			201	Harris
			221	Hood
			231	Hunt
			251	Johnson
			277	Lamar

State Code	State	# of Cases per county (Blank-1)	County Code	County
42	Texas		303	Lubbock
			323	Maverick
			343	Morris
			369	Parmer
			399	Runnels
			439	Tarrant
			463	Uvalde
			485	Wichita
43	Utah		057	Weber
45	Virginia		009	Amherst
			153	Prince William
			670	Hopewell
46	Washington		009	Clallam
			025	Grant
			033	King
			037	Kittitas
			049	Pacific
			057	Skagit
			063	Spokane
			073	Whatcom
47	West Virginia		005	Boone
			029	Hancock
			039	Kanawha
			053	Mason
			069	Ohio
			083	Randolph
			105	Wirt
49	Wyoming		013	Fremont
			033	Sherican
51	Hawaii		003	Honolulu
			009	Maui

LITERATURE CITED

Black, Ralph J., et.al. The National Solid Waste Survey: An
    Interim Report 2nd Printing, (Washington: U. S. Environmental
    Protection Agency 1970).

California Solid Waste Management Study 1968 and Plan 1970,
    (Washington: U. S. Environmental Protection Agency 1971).

Coding Manual: The National Survey of Community Solid Waste
    Practices, (Washington: Dept. of HEW, Solid Waste Program,
    U. S. Public Health Service, September, 1967).

County and City Data Book 1972: A Statistical Abstract
    Supplement, (Washington: U. S. Bureau of Census, March, 1973).

Davidson, George R., Residential Solid Waste Generation in Low
    Income Areas, (Washington: U. S. Environmental Protection
    Agency, 1972).

Geographical Location Codes, (Washington: U.S.G.P.O., General
    Administrations Office of Finance, October, 1966).

Jarvis, Rich, SELECT: A Utility Program to Manipulate a
    Sequential File, (Chicago: National Opinion Research Center,
    U. of Chicago, September, 1972).

Klee, Albert J., "Mapping the United States - A Solid Waste
    View." Waste Age, September-October, 1970.

Knotuly, Thomas, Solid Waste Generation Coefficient: Non-
    Manufacturing Sectors, Regional Science Research Institute,
    Working Paper (Philadelphia, April, 1974).

Muhich, A. J., "Sample Representativeness and Community Data."
    The 1968 National Survey of Community Solid Waste Practices:
    An Interim Report, (Washington: U. S. Dept. HEW, 1970).

Muhich, A. J., Klee, A. J., Hampel, Charles R., The 1968 National
    Survey of Community Solid Waste Practices Regions 1 & 2,
    (Washington: U. S. Dept. of HEW, Public Health Service,
    1969).

Muhich, A. J., Klee, A. J. and Brittons, P. W., Preliminary
    Data Analysis: 1968 National Survey of Community Solid
    Waste Practices, (Cincinnati: Dept. of HEW Public Health
    Service, Publication #1867, 1968).

Municipal Refuse Disposal 3rd ed. (Chicago: American Public Works Association 1970).

Nie, Norman H., Hull, C. Hadlai, et. al., Statistical Package for the Social Sciences (SPSS), 2nd ed. (New York: McGraw-Hill Book Co., Inc., 1970).

Proceeding National Conference on Solid Waste Research, (Chicago: Research Foundation, American Public Works Association, 1963).

Refuse Collection Practices 3rd ed., (Chicago: American Public Works Association, 1966).

Solid Waste Management Glossary (Washington: U. S. Environmental Protection Agency, 1972).

Steiker, Gene, Solid Waste Generation Coefficients: Manufacturing Sectors, Regional Science Research Institute Discussion Paper Paper 70 (Philadelphis; December, 1973).

Weston, Ray F., New York Solid Waste Management Plan, Status Report 1970, (Washington: U. S. Environmental Protection Agency, 1971).

THE UNIVERSITY OF CHICAGO
DEPARTMENT OF GEOGRAPHY
RESEARCH PAPERS (Lithographed, 6×9 Inches)

(Available from Department of Geography, The University of Chicago, 5828 S. University Ave., Chicago, Illinois 60637. Price: $6.00 each; by series subscription, $5.00 each.)

106. SAARINEN, THOMAS F. *Perception of the Drought Hazard on the Great Plains* 1966. 183 pp.

107. SOLZMAN, DAVID M. *Waterway Industrial Sites: A Chicago Case Study* 1967. 138 pp.

108. KASPERSON, ROGER E. *The Dodecanese: Diversity and Unity in Island Politics* 1967. 184 pp.

109. LOWENTHAL, DAVID, et al. *Environmental Perception and Behavior.* 1967. 88 pp.

110. REED, WALLACE E. *Areal Interaction in India: Commodity Flows of the Bengal-Bihar Industrial Area* 1967. 210 pp.

112. BOURNE, LARRY S. *Private Redevelopment of the Central City: Spatial Processes of Structural Change in the City of Toronto* 1967. 199 pp.

113. BRUSH, JOHN E., and GAUTHIER, HOWARD L., JR. *Service Centers and Consumer Trips: Studies on the Philadelphia Metropolitan Fringe* 1968. 182 pp.

114. CLARKSON, JAMES D. *The Cultural Ecology of a Chinese Village: Cameron Highlands, Malaysia* 1968. 174 pp.

115. BURTON, IAN; KATES, ROBERT W.; and SNEAD, RODMAN E. *The Human Ecology of Coastal Flood Hazard in Megalopolis* 1968. 196 pp.

117. WONG, SHUE TUCK. *Perception of Choice and Factors Affecting Industrial Water Supply Decisions in Northeastern Illinois* 1968. 96 pp.

118. JOHNSON, DOUGLAS L. *The Nature of Nomadism* 1969. 200 pp.

119. DIENES, LESLIE. *Locational Factors and Locational Developments in the Soviet Chemical Industry* 1969. 285 pp.

120. MIHELIC, DUSAN. *The Political Element in the Port Geography of Trieste* 1969. 104 pp.

121. BAUMANN, DUANE. *The Recreational Use of Domestic Water Supply Reservoirs: Perception and Choice* 1969. 125 pp.

122. LIND, AULIS O. *Coastal Landforms of Cat Island, Bahamas: A Study of Holocene Accretionary Topography and Sea-Level Change* 1969. 156 pp.

123. WHITNEY, JOSEPH. *China: Area, Administration and Nation Building* 1970. 198 pp.

124. EARICKSON, ROBERT. *The Spatial Behavior of Hospital Patients: A Behavioral Approach to Spatial Interaction in Metropolitan Chicago* 1970. 198 pp.

125. DAY, JOHN C. *Managing the Lower Rio Grande: An Experience in International River Development* 1970. 277 pp.

126. MAC IVER, IAN. *Urban Water Supply Alternatives: Perception and Choice in the Grand Basin, Ontario* 1970. 178 pp.

127. GOHEEN, PETER G. *Victorian Toronto, 1850 to 1900: Pattern and Process of Growth* 1970. 278 pp.

128. GOOD, CHARLES M. *Rural Markets and Trade in East Africa* 1970. 252 pp.

129. MEYER, DAVID R. *Spatial Variation of Black Urban Households* 1970. 127 pp.

130. GLADFELTER, BRUCE. *Meseta and Campiña Landforms in Central Spain: A Geomorphology of the Alto Henares Basin* 1971. 204 pp.

131. NEILS, ELAINE M. *Reservation to City: Indian Urbanization and Federal Relocation* 1971. 200 pp.

132. MOLINE, NORMAN T. *Mobility and the Small Town, 1900–1930* 1971. 169 pp.

133. SCHWIND, PAUL J. *Migration and Regional Development in the United States, 1950–1960* 1971. 170 pp.

134. PYLE, GERALD F. *Heart Disease, Cancer and Stroke in Chicago: A Geographical Analysis with Facilities Plans for 1980* 1971. 292 pp.

135. JOHNSON, JAMES F. *Renovated Waste Water: An Alternative Source of Municipal Water Supply in the U.S.* 1971. 155 pp.

136. BUTZER, KARL W. *Recent History of an Ethiopian Delta: The Omo River and the Level of Lake Rudolf* 1971. 184 pp.

137. HARRIS, CHAUNCY D. *Annotated World List of Selected Current Geographical Serials in English, French, and German* 3rd edition 1971. 77 pp.

138. HARRIS, CHAUNCY D., and FELLMANN, JEROME D. *International List of Geographical Serials* 2nd edition 1971. 267 pp.

139. MC MANIS, DOUGLAS R. *European Impressions of the New England Coast, 1497–1620* 1972. 147 pp.

140. COHEN, YEHOSHUA S. *Diffusion of an Innovation in an Urban System: The Spread of Planned Regional Shopping Centers in the United States, 1949–1968* 1972. 136 pp.

141. MITCHELL, NORA. *The Indian Hill-Station: Kodaikanal* 1972. 199 pp.

142. PLATT, RUTHERFORD H. *The Open Space Decision Process: Spatial Allocation of Costs and Benefits* 1972. 189 pp.

143. GOLANT, STEPHEN M. *The Residential Location and Spatial Behavior of the Elderly: A Canadian Example* 1972. 226 pp.

144. PANNELL, CLIFTON W. *T'ai-chung, T'ai-wan: Structure and Function* 1973. 200 pp.

145. LANKFORD, PHILIP M. *Regional Incomes in the United States, 1929–1967: Level, Distribution, Stability, and Growth* 1972. 137 pp.

146. FREEMAN, DONALD B. *International Trade, Migration, and Capital Flows: A Quantitative Analysis of Spatial Economic Interaction* 1973. 202 pp.

147. MYERS. SARAH K. *Language Shift Among Migrants to Lima, Peru* 1973. 204 pp.

148. JOHNSON, DOUGLAS L. *Jabal al-Akhdar, Cyrenaica: An Historical Geography of Settlement and Livelihood* 1973. 240 pp.

149. YEUNG, YUE-MAN. *National Development Policy and Urban Transformation in Singapore: A Study of Public Housing and the Marketing System* 1973. 204 pp.

150. HALL, FRED L. *Location Criteria for High Schools: Student Transportation and Racial Integration* 1973. 156 pp.

151. ROSENBERG, TERRY J. *Residence, Employment, and Mobility of Puerto Ricans in New York City* 1974. 230 pp.

152. MIKESELL, MARVIN W., editor. *Geographers Abroad: Essays on the Problems and Prospects of Research in Foreign Areas* 1973. 296 pp.

153. OSBORN, JAMES. *Area, Development Policy, and the Middle City in Malaysia* 1974. 273 pp.

154. WACHT, WALTER F. *The Domestic Air Transportation Network of the United States* 1974. 98 pp.

155. BERRY, BRIAN J. L., et al. *Land Use, Urban Form and Environmental Quality* 1974. 464 pp.

156. MITCHELL, JAMES K. *Community Response to Coastal Erosion: Individual and Collective Adjustments to Hazard on the Atlantic Shore* 1974. 209 pp.

157. COOK, GILLIAN P. *Spatial Dynamics of Business Growth in the Witwatersrand* 1975. 143 pp.

158. STARR, JOHN T., JR. *The Evolution of Unit Train Operations in the United States: 1960–1969—A Decade of Experience* 1976. 247 pp.

159. PYLE, GERALD F. *The Spatial Dynamics of Crime* 1974. 220 pp.

160. MEYER, JUDITH W. *Diffusion of an American Montessori Education* 1975. 109 pp.

161. SCHMID, JAMES A. *Urban Vegetation: A Review and Chicago Case Study* 1975. 280 pp.

162. LAMB, RICHARD. *Metropolitan Impacts on Rural America* 1975. 210 pp.

163. FEDOR, THOMAS. *Patterns of Urban Growth in the Russian Empire during the Nineteenth Century* 1975. 275 pp.

164. HARRIS, CHAUNCY D. *Guide to Geographical Bibliographies and Reference Works in Russian or on the Soviet Union* 1975. 496 pp.

165. JONES, DONALD W. *Migration and Urban Unemployment in Dualistic Economic Development* 1975. 186 pp.

166. BEDNARZ, ROBERT S. *The Effect of Air Pollution on Property Value in Chicago* 1975. 118 pp.

167. HANNEMANN, MANFRED. *The Diffusion of the Reformation in Southwestern Germany, 1518-1534* 1975. 248 pp.

168. SUBLETT, MICHAEL D. *Farmers on the Road. Interfarm Migration and the Farming of Noncontiguous Lands in Three Midwestern Townships, 1939-1969* 1975. 228 pp.

169. STETZER, DONALD FOSTER. *Special Districts in Cook County: Toward a Geography of Local Government* 1975. 189 pp.

170. EARLE, CARVILLE V. *The Evolution of a Tidewater Settlement System: All Hallow's Parish, Maryland, 1650-1783* 1975. 249 pp.

171. SPODEK, HOWARD. *Urban-Rural Integration in Regional Development: A Case Study of Saurashtra, India—1800-1960* 1976. 156 pp.

172. COHEN, YEHOSHUA S. and BERRY, BRIAN J. L. *Spatial Components of Manufacturing Change* 1975 272 pp.

173. HAYES, CHARLES R. *The Dispersed City: The Case of Piedmont, North Carolina* 1976. 169 pp.

174. CARGO, DOUGLAS B. *Solid Wastes: Factors Influencing Generation Rates* 1977. 111 pp.

175. GILLARD, QUENTIN. *Incomes and Accessibility. Metropolitan Labor Force Participation, Commuting, and Income Differentials in the United States, 1960-1970* 1977. 140 pp.

176. MORGAN, DAVID J. *Patterns of Population Distribution: A Residential Preference Model and Its Dynamic* 1977.

177. STOKES, HOUSTON H.; JONES, DONALD W. and NEUBURGER, HUGH M. *Unemployment and Adjustment in the Labor Market: A Comparison between the Regional and National Responses* 1975 135 pp.

178. PICCAGLI, GIORGIO ANTONIO. *Racial Transition in Chicago Public Schools. An Examination of the Tipping Point Hypothesis, 1963–1971* 1977.

179. HARRIS, CHAUNCY D. *Bibliography of Geography. Part I. Introduction to General Aids* 1976. 288 pp.

180. CARR, CLAUDIA J. *Pastoralism in Crisis. The Dasanetch and their Ethiopian Lands.* 1977. 339 pp.

181. GOODWIN, GARY C. *Cherokees in Transition: A Study of Changing Culture and Environment Prior to 1775.* 1977. 221 pp.

182. KNIGHT, DAVID B. *A Capital for Canada: Conflict and Compromise in the Nineteenth Century.* 1977. 359 pp.

183. HAIGH, MARTIN J. *The Evolution of Slopes on Artificial Landforms: Blaenavon, Gwent.* 1978. 311 pp.

184. FINK, L. DEE. *Listening to the Learner. An Exploratory Study of Personal Meaning in College Geography Courses.* 1977. 200 pp.

185. HELGREN, DAVID M. *Rivers of Diamonds: An Alluvial History of the Lower Vaal Basin.* 1978.

186. BUTZER, KARL W., *editor. Dimensions of Human Geography: Essays on Some Familiar and Neglected Themes.* 1978. 201 pp.

187. MITSUHASHI, SETSUKO. *Japanese Commodity Flows.* 1978. 185 pp.